This report contains the collective views of an international group of experts and does not necessarily represent the decisions or the stated policy of the United Nations Environment Programme, the International Labour Organisation, or the World Health Organization.

Environmental Health Criteria 220

DINITRO-*ortho*-CRESOL

First draft prepared by Dr A.F. Pelfrène, Charbonnières-les-Bains, France

Published under the joint sponsorship of the United Nations Environment Programme, the International Labour Organisation, and the World Health Organization, and produced within the framework of the Inter-Organization Programme for the Sound Management of Chemicals.

World Health Organization
Geneva, 2000

The **International Programme on Chemical Safety** (IPCS), established in 1980, is a joint venture of the United Nations Environment Programme (UNEP), the International Labour Organization (ILO), and the World Health Organization (WHO). The overall objectives of the IPCS are to establish the scientific basis for assessment of the risk to human health and the environment from exposure to chemicals, through international peer-review processes, as a prerequisite for the promotion of chemical safety, and to provide technical assistance in strengthening national capacities for the sound management of chemicals.

The **Inter-Organization Programme for the Sound Management of Chemicals** (IOMC) was established in 1995 by UNEP, ILO, the Food and Agriculture Organization of the United Nations, WHO, the United Nations Industrial Development Organization, the United Nations Institute for Training and Research, and the Organisation for Economic Co-operation and Development (Participating Organizations), following recommendations made by the 1992 UN Conference on Environment and Development to strengthen cooperation and increase coordination in the field of chemical safety. The purpose of the IOMC is to promote coordination of the policies and activities pursued by the Participating Organizations, jointly or separately, to achieve the sound management of chemicals in relation to human health and the environment.

WHO Library Cataloguing-in-Publication Data

Dinitro-*ortho*-cresol.

(Environmental health criteria ; 220)

1. Cresols - chemistry 2. Cresols - toxicity 3. Dinitrophenols - chemistry 4. Dinitrophenols - toxicity 5. Occupational exposure 6. Environmental exposure 7. Risk assessment I. Series

ISBN 92 4 157220 5 (NLM Classification: QD 341.P5)
ISSN 0250-863X

The World Health Organization welcomes requests for permission to reproduce or translate its publications, in part or in full. Applications and enquiries should be addressed to the Office of Publications, World Health Organization, Geneva, Switzerland, which will be glad to provide the latest information on any changes made to the text, plans for new editions, and reprints and translations already available.

©World Health Organization 2000

Publications of the World Health Organization enjoy copyright protection in accordance with the provisions of Protocol 2 of the Universal Copyright Convention. All rights reserved.

The designations employed and the presentation of the material in this publication do not imply the expression of any opinion whatsoever on the part of the Secretariat of the World Health Organization concerning the legal status of any country, territory, city or area or of its authorities, or concerning the delimitation of its frontiers or boundaries.

The mention of specific companies or of certain manufacturers' products does not imply that they are endorsed or recommended by the World Health Organization in preference to others of a similar nature that are not mentioned. Errors and omissions excepted, the names of proprietary products are distinguished by initial capital letters.

Edited and typeset by Kathleen Lyle Editorial Services, Sheffield, England

The Federal Ministry for the Environment, Nature Conservation and Nuclear Safety, Germany, provided financial support for, and undertook the printing of, this publication.

Printed by Wissenschaftliche Verlagsgesellschaft mbH, D-70009 Stuttgart 10

CONTENTS

1. SUMMARY AND CONCLUSIONS   1

   1.1   Summary   1
         1.1.1   Identity, physical and chemical properties
                 and analytical methods   1
         1.1.2   Sources of human and environmental exposure   1
         1.1.3   Environmental transport, distribution
                 and transformation   1
         1.1.4   Environmental levels and human exposure   2
         1.1.5   Kinetics and metabolism   2
         1.1.6   Effects on laboratory mammals;
                 in vitro test systems   2
                 1.1.6.1   Single exposure   2
                 1.1.6.2   Short-term exposure   2
                 1.1.6.3   Skin and eye irritation and skin
                           sensitization   3
                 1.1.6.4   Long-term exposure   3
                 1.1.6.5   Reproduction, embryotoxicity
                           and teratogenicity   3
                 1.1.6.6   Mutagenicity   3
                 1.1.6.7   Carcinogenicity   4
         1.1.7   Effects on humans   4
         1.1.8   Effects on organisms in the environment   4
   1.2   Conclusions   4

2. IDENTITY, PHYSICAL AND CHEMICAL PROPERTIES,
   ANALYTICAL METHODS   6

   2.1   Chemical identity   6
   2.2   Physical and chemical properties   8
   2.3   Analytical methods   8

3. SOURCES OF HUMAN AND ENVIRONMENTAL
   EXPOSURE   15

   3.1   Natural occurrence   15
   3.2   Anthropogenic sources   15
         3.2.1   Uses   15
         3.2.2   Worldwide sales   16

4. ENVIRONMENTAL TRANSPORT, DISTRIBUTION AND TRANSFORMATION 17

 4.1 Transport and distribution between media 17
  4.1.1 Air 17
  4.1.2 Water 17
  4.1.3 Soil 17
 4.2 Degradation 18
 4.3 Crop uptake 19

5. ENVIRONMENTAL LEVELS AND HUMAN EXPOSURE 20

 5.1 Environmental levels 20
  5.1.1 Air 20
  5.1.2 Water and soil 20
  5.1.3 Food and feed 21
 5.2 General population exposure 21
  5.2.1 Oral exposure 21
  5.2.2 Inhalation exposure 21
 5.3 Occupational exposure during manufacturing, formulation and use 21

6. KINETICS AND METABOLISM 24

 6.1 Absorption 24
 6.2 Distribution and accumulation 25
 6.3 Biotransformation 26
 6.4 Elimination and excretion 28
 6.5 Reaction with body components 29

7. EFFECTS ON LABORATORY MAMMALS; *IN VITRO* TEST SYSTEMS 30

 7.1 Single exposure 30
  7.1.1 Oral exposure 30
  7.1.2 Inhalation exposure 30
  7.1.3 Skin exposure 30
  7.1.4 Skin sensitization 31
 7.2 Short-term exposure 32
  7.2.1 Oral administration 32
   7.2.1.1 Rat 32
   7.2.1.2 Mouse 34

	7.2.1.3 Dog	34
7.2.2	Inhalation	35
	7.2.2.1 Cat	35
7.3	Skin and eye irritation; skin sensitization	35
7.4	Long-term exposure	35
7.4.1	Rat	35
7.5	Reproduction, embryotoxicity and teratogenicity	36
7.5.1	Reproduction	36
7.5.2	Teratogenicity and embryotoxicity	37
	7.5.2.1 Rat oral study	37
	7.5.2.2 Mouse oral study	37
	7.5.2.3 Rabbit oral studies	38
	7.5.2.4 Rabbit dermal studies	38
	7.5.2.5 Mouse intraperitoneal studies	39
7.6	Mutagenicity and related endpoints	39
7.6.1	Microbial systems and lower organisms	44
7.6.2	Mammalian cells *in vitro*	44
7.6.3	Mammalian cells *in vivo*	44
7.7	Carcinogenicity	45
7.8	Special studies	45
7.8.1	Cataractogenicity	45
7.8.2	Immunotoxicity	45
7.8.3	Testicular toxicity	46
7.9	Factors modifying toxicity; toxicity of metabolites	46
7.9.1	Factors modifying toxicity	46
7.9.2	Toxicity of metabolites	47
7.10	Mechanisms of toxicity; mode of action	47

8. EFFECTS ON HUMANS 49

8.1	General population exposure	49
8.1.1	Clinical studies	49
8.1.2	Acute toxicity	49
8.2	Occupational exposure	49

9. EFFECTS ON ORGANISMS IN THE LABORATORY AND FIELD 55

9.1	Micro-organisms	55
9.2	Aquatic organisms	55
9.3	Terrestrial organisms	55
9.3.1	Earthworms	55
9.3.2	Honey bees	55

9.3.3	Birds	58
9.3.4	Other wildlife species	59

10. EVALUATION OF HUMAN HEALTH RISKS AND EFFECTS ON THE ENVIRONMENT — 60

    10.1 Evaluation of human health risks — 60
    10.2 Evaluation of effects on the environment — 61

11. PREVIOUS EVALUATION BY INTERNATIONAL BODIES — 63

REFERENCES — 64

Résumé et conclusions — 76
Resumen y conclusiones — 81

NOTE TO READERS OF THE CRITERIA MONOGRAPHS

Every effort has been made to present information in the criteria monographs as accurately as possible without unduly delaying their publication. In the interest of all users of the Environmental Health Criteria monographs, readers are requested to communicate any errors that may have occurred to the Director of the International Programme on Chemical Safety, World Health Organization, Geneva, Switzerland, in order that they may be included in corrigenda.

A detailed data profile and a legal file can be obtained from the International Register of Potentially Toxic Chemicals, Case postale 356, 1219 Chatelaine, Geneva, Switzerland (telephone no. + 41 22 9799111, fax no. + 41 22 7973460, E-mail irptcp@unep.ch).

This publication was made possible by grant number 5 U01 ES02617-15 from the National Institute of Environmental Health Sciences, National Institutes of Health, USA, and by financial support from the European Commission.

Environmental Health Criteria
PREAMBLE

Objectives

In 1973 the WHO Environmental Health Criteria Programme was initiated with the following objectives:

- to assess information on the relationship between exposure to environmental pollutants and human health, and to provide guidelines for setting exposure limits;
- to identify new or potential pollutants;
- to identify gaps in knowledge concerning the health effects of pollutants;
- to promote the harmonization of toxicological and epidemiological methods in order to have internationally comparable results.

The first Environmental Health Criteria (EHC) monograph, on mercury, was published in 1976 and since that time an ever-increasing number of assessments of chemicals and of physical effects have been produced. In addition, many EHC monographs have been devoted to evaluating toxicological methodology, e.g., for genetic, neurotoxic, teratogenic and nephrotoxic effects. Other publications have been concerned with epidemiological guidelines, evaluation of short-term tests for carcinogens, biomarkers, effects on the elderly and so forth.

Since its inauguration the EHC Programme has widened its scope, and the importance of environmental effects, in addition to health effects, has been increasingly emphasized in the total evaluation of chemicals.

The original impetus for the Programme came from World Health Assembly resolutions and the recommendations of the 1972 UN Conference on the Human Environment. Subsequently the work became an integral part of the International Programme on Chemical Safety (IPCS), a cooperative programme of UNEP, ILO and WHO. In this manner, with the strong support of the new partners, the importance of occupational health and environmental effects was fully recognized. The EHC monographs have become widely established, used and recognized throughout the world.

The recommendations of the 1992 LTN Conference on Environment and Development and the subsequent establishment of the Intergovernmental Forum on Chemical Safety with the priorities for action in the six programme areas of Chapter 19, Agenda 2 1, all lend further weight to the need for EHC assessments of the risks of chemicals.

Scope

The criteria monographs are intended to provide critical reviews on the effect on human health and the environment of chemicals and of combinations of chemicals and physical and biological agents. As such, they include and review studies that are of direct relevance for the evaluation. However, they do not describe every study carried out. Worldwide data are used and are quoted from original studies, not from abstracts or reviews. Both published and unpublished reports are considered and it is incumbent on the authors to assess all the articles cited in the references. Preference is always given to published data. Unpublished data are used only when relevant published data are absent or when they are pivotal to the risk assessment. A detailed policy statement is available that describes the procedures used for unpublished proprietary data so that this information can be used in the evaluation without compromising its confidential nature (WHO (1990) Revised Guidelines for the Preparation of Environmental Health Criteria Monographs. PCS/90.69, Geneva, World Health Organization).

In the evaluation of human health risks, sound human data, whenever available, are preferred to animal data. Animal and *in vitro* studies provide support and are used mainly to supply evidence missing from human studies. It is mandatory that research on human subjects is conducted in full accord with ethical principles, including the provisions of the Helsinki Declaration.

The EHC monographs are intended to assist national and international authorities in making risk assessments and subsequent risk management decisions. They represent a thorough evaluation of risks and are not, in any sense, recommendations for regulation or standard setting. These latter are the exclusive purview of national and regional governments.

Content

The layout of EHC monographs for chemicals is outlined below.

- Summary – a review of the salient facts and the risk evaluation of the chemical
- Identity – physical and chemical properties, analytical methods
- Sources of exposure
- Environmental transport, distribution and transformation
- Environmental levels and human exposure
- Kinetics and metabolism in laboratory animals and humans
- Effects on laboratory manuals and *in vitro* test systems
- Effects on humans
- Effects on other organisms in the laboratory and field
- Evaluation of human health risks and effects on the environment
- Conclusions and recommendations for protection of human health and the environment
- Further research
- Previous evaluations by international bodies, e.g., LARC, TECFA, JMPR

Selection of chemicals

Since the inception of the EHC Programme, the IPCS has organized meetings of scientists to establish lists of priority chemicals for subsequent evaluation. Such meetings have been held in: Ispra, Italy, 1980; Oxford, United Kingdom, 1984; Berlin, Germany, 1987; and North Carolina, USA, 1995. The selection of chemicals has been based on the following criteria: the existence of scientific evidence that the substance presents a hazard to human health and/or the environment; the possible use, persistence, accumulation or degradation of the substance shows that there may be significant human or environmental exposure; the size and nature of populations at risk (both human and other species) and risks for the environment; international concern, i.e. the substance is of major interest to several countries; adequate data on the hazards are available.

If an EHC monograph is proposed for a chemical not on the priority list, the IPCS Secretariat consults with the Cooperating Organizations and all the Participating Institutions before embarking on the preparation of the monograph.

EHC PREPARATION FLOW CHART

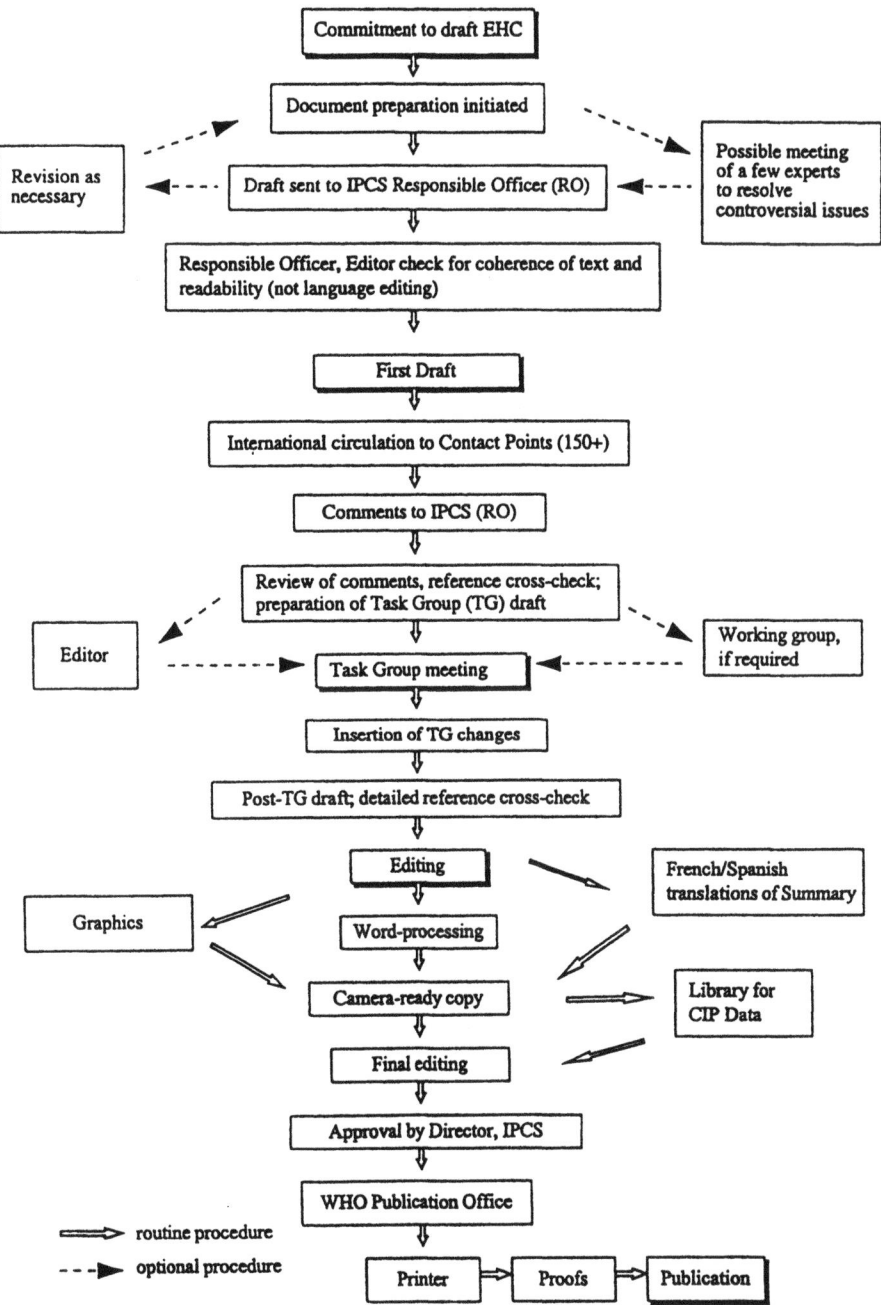

Procedures

The order of procedures that result in the publication of an EHC monograph is shown in the flow chart. A designated staff member of IPCS, responsible for the scientific quality of the document, serves as Responsible Officer (RO). The IPCS Editor is responsible for layout and language. The first draft, prepared by consultants or, more usually, staff from an IPCS Participating Institution, is based initially on data provided from the International Register of Potentially Toxic Chemicals, and reference data bases such as Medline and Toxline.

The draft document, when received by the RO, may require an initial review by a small panel of experts to determine its scientific quality and objectivity. Once the RO finds the document acceptable as a first draft, it is distributed, in its unedited form, to well over 150 EHC contact points throughout the world who are asked to comment on its completeness and accuracy and, where necessary, provide additional material. The contact points, usually designated by governments, may be Participating Institutions, IPCS Focal Points, or individual scientists known for their particular expertise. Generally some four months are allowed before the comments are considered by the RO and author(s). A second draft incorporating comments received and approved by the Director, IPCS, is then distributed to Task Group members, who carry out the peer review, at least six weeks before their meeting.

The Task Group members serve as individual scientists, not as representatives of any organization, government or industry. Their function is to evaluate the accuracy, significance and relevance of the information in the document and to assess the health and environmental risks from exposure to the chemical. A summary and recommendations for further research and improved safety aspects are also required. The composition of the Task Group is dictated by the range of expertise required for the subject of the meeting and by the need for a balanced geographical distribution.

The three cooperating organizations of the IPCS recognize the important role played by non-govermental organizations. Representatives from relevant national and international associations may be invited to join the Task Group as observers. While observers may provide a valuable contribution to the process, they can speak only at the invitation of the Chairperson. Observers do not participate in the final evaluation of the chemical; this is the sole responsibility of the Task Group members. When the Task Group considers it to be appropriate, it may meet *in camera*.

All individuals who as authors, consultants or advisers participate in the preparation of the EHC monograph must, in addition to serving in their personal capacity as scientists, inform the RO if at any time a conflict of interest, whether actual or potential, could be perceived in their work. They are required to sign a conflict of interest statement. Such a procedure ensures the transparency and probity of the process.

When the Task Group has completed its review and the RO is satisfied as to the scientific correctness and completeness of the document, it then goes for language editing, reference checking, and preparation of camera-ready copy. After approval by the Director, IPCS, the monograph is submitted to the VMO Office of Publications for printing. At this time a copy of the final draft is sent to the Chairperson and Rapporteur of the Task Group to check for any errors.

It is accepted that the following criteria should initiate the updating of an EHC monograph: new data are available that would substantially change the evaluation; there is public concern for health or environmental effects of the agent because of greater exposure; an appreciable time period has elapsed since the last evaluation.

All Participating Institutions are informed, through the EHC progress report, of the authors and institutions proposed for the drafting of the documents. A comprehensive file of all comments received on drafts of each EHC monograph is maintained and is available on request. The Chairpersons of Task Groups are briefed before each meeting on their role and responsibility in ensuring that these rules are followed.

IPCS TASK GROUP ON ENVIRONMENTAL HEALTH CRITERIA FOR DINITRO-*ortho*-CRESOL

Members

Dr D. Anderson, British Industrial Biological Research Association (BIBRA) International, Carshalton, Surrey, United Kingdom

Dr B.H. Chen, Department of Environmental Health, School of Public Health, Shanghai Medical University, Shanghai, People's Republic of China

Dr S. Dobson, The Institute of Terrestrial Ecology, Monks Wood Experimental Station, Abbots Ripton, Huntingdon, Cambridgeshire, United Kingdom

Professor M.C.A. Lotti, Università degli Studi di Padova, Istituto di Medicina del Lavoro, Azienda Ospedaliera, Padova, Italy (*Chairman*)

Dr P. Lundberg, Risk Evaluation Group, Department of Occupational Medicine, National Institute for Working Life, Solna, Sweden

Dr L.R. Papa, National Center for Environmental Assessment – CIN, US Environmental Protection Agency, Cincinnati, Ohio, USA

Dr A.F. Pelfrène, The Agrochemicals Defense Network, La Marjolaine, Charbonnières-les-Bains, France (*Rapporteur*)

Professor S.A. Soliman, Department of Pesticide Chemistry, Faculty of Agriculture, Alexandria University, El-Shatby, Alexandria, Egypt

Secretariat

Mr Y. Hayashi, Scientist, International Programme on Chemical Safety, World Health Organization, Geneva, Switzerland

Dr Y. Uyama, Food Chemistry Division, Environmental Health Bureau, Ministry of Health and Welfare, Tokyo, Japan (On secondment to the International Programme on Chemical Safety)

Dr M. Younes, Acting Coordinator, International Programme on Chemical Safety, World Health Organization, Geneva, Switzerland (*Secretary*)

WHO TASK GROUP ON ENVIRONMENTAL HEALTH CRITERIA FOR DINITRO-*ortho*-CRESOL

A WHO Task Group on Environmental Health Criteria for Dinitro-*ortho*-cresol was held at the World Health Organization, Geneva, Switzerland from 20 to 23 April 1999. Dr R. Helmer, Director, Department for the Protection of the Human Environment, opened the meeting and welcomed the participants on behalf the IPCS and its three cooperating organizations (UNEP/ILO/WHO). The Task Group reviewed and revised the draft criteria monograph and made an evaluation of the risks for human health and the environment from exposure to dinitro-*ortho*-cresol.

Dr A.F. Pelfrène prepared the first draft of this monograph. The second draft incorporated comments received following the circulation of the first draft to the IPCS Contact Points for Environmental Health Criteria monographs.

Dr B.H. Chen (IPCS) and Ms K. Lyle (Sheffield, England) were responsible for the overall scientific content and technical editing, respectively.

The efforts of all who helped in the preparation and finalization of the monograph are gratefully acknowledged.

* * *

Financial support for this Task Group was provided by the US Food and Drug Administration as part of its contributions to the IPCS.

ABBREVIATIONS

4-ANOC	4-amino-6-nitro-*o*-cresol
6-ANOC	6-amino-4-nitro-*o*-cresol
6-Ac ANOC	6-acetamido-4-nitro-*o*-cresol
ADI	acceptable daily intake
ADP	adenosine disphosphate
AdSV	adsorptive stripping voltametric detector
a.i.	active ingredient
ALT	alanine aminotrasferase
3-ANSA	3-amino-5-nitrosalicyclic acid
AST	aspartate aminotransferase
ATP	adenosine triphosphate
b.w.	body weight
BMR	basal metabolic rate
BOEL	biological operation exposure limit
BSI	British Standards Institute
CA	Chemical Abstracts
CAS	Chemical Abstracts Services
DECOS	Dutch Expert Committee on Occupational Standards
DNC	synonym for DNOC
DNHMP	4,6-dinitro-2-hydroxymethylphenol
DNOC	4,6 dinitro-*o*-cresol
DPP	differential pulse polarographic detector
DT_{50}	median degradation time
EC	emulsifiable concentrate
EC_{50}	median effective concentration
ELCD	electrochemical detector
ENT 154	synonym for DNOC
EPPO	European and Mediterranean Plant Protection Organization
FID	flame ionization detection
F_0	first filial generation
GC	gas chromatography
GLP	Good Laboratory Practice
GTZ	German Agency for Technical Cooperation
HPLC	high-performance liquid chromatography
HRGC	high-resolution gas chromatography
ISO	International Organization for Standardization
IUPAC	International Union of Pure and Applied Chemistry
JMAF	Japanese Ministry of Agriculture and Forestry
JMPR	FAO/WHO Joint Meeting on Pesticide Residues
LC_{50}	median lethal concentration

LC–MS	liquid chromatography–mass spectrometry
LD_{50}	median lethal dose
MAC_{WZ}	maximum allowable concentration in the working zone
MRL	maximum residue limit
MS	mass spectrometry
MS–MS	tandem mass spectrometry
MTD	maximum tolerated dose
NOAEL	no observed adverse effect level
NOEC	no observed effect concentration
NOEL	no effect level
NPD	nitrogen phosphorus detector
OECD	Organisation for Economic Co-operation and Development
OL	oil-miscible liquids
PA	pastes
PDD	photodiode array detector
PND	phosphorus/nitrogen detector
PT_{50}	median photolysis time
RSD	relative standard deviation
SC	suspension concentrate
SGOT	*see* ALT
SGPT	*see* AST
SPE	solid phase extraction
SPME	solid phase microextraction
$t_{1/2}$	half-life
T_3	triiodothyronine
T_4	thyroxine
JMPR	FAO/WHO Joint Meeting on Pesticide Residues
TER	toxicity exposure ratio
$TSEL_{hm}$	tentatively safe exposure level in the atmosphere of residential areas
TWA	time weighted average
UV	ultraviolet
v/v	volume per volume

1. SUMMARY AND CONCLUSIONS

1.1 Summary

1.1.1 Identity, physical and chemical properties and analytical methods

DNOC (4-6 dinitro-*ortho*-cresol) is a yellowish crystalline solid. Its melting point is 88.2–88.9 °C and its vapour pressure is 1.6×10^{-2} Pa at 25 °C.

The solubility of DNOC in water is 6.94 g/litre at 20 °C and pH 7, and largely depends on pH.

DNOC is relatively stable in sterile water.

DNOC is analysed in environmental media by high-performance liqid chromatrography (HPLC) with ultraviolet (UV) detection or by gas chromatography (GC) with detection by nitrogen phosphorus dection (NPD), flame ionization detection (FID) or mass spectrometry (MS). In biological fluids, determination of DNOC is usually by spectrophotometry and more recently by either GC/NPD or HPLC/UV.

1.1.2 Sources of human and environmental exposure

DNOC is used agriculturally as a larvicide, ovicide and insecticide (against locusts and other insects) as well as a potato haulm desiccant. It is also used as a polymerization inhibitor and as an intermediate in the chemical industry. For agricultural uses, DNOC is mainly formulated as emulsifiable concentrate, either aqueous or oily.

1.1.3 Environmental transport, distribution and transformation

The half-life of DNOC in surface water is 3–5 weeks. Its low vapour pressure and moderate water solubility indicate that DNOC has no potential to volatilize. In soils, DNOC is rapidly degraded by microorganisms with median degradation time (DT_{50}) values in the range of 1.7–15 days. Several environmental metabolites have been identified, resulting from a reductive biotransformation possibly followed by further oxidative degradation. Adsorption of undissociated DNOC to particulates is strong at low pH, but sorption is limited at environmentally relevant pH. In practice, little leaching to groundwater has been found, probably because of biodegradation.

1.1.4 Environmental levels and human exposure

The main sources of human exposure are from contact during manufacturing, and from use in agriculture and in the plastics industry. Because of the known acute toxicity and the strong yellow staining of the skin, agricultural workers are careful to use adequate protective clothing in order to reduce dermal exposure. In the plastics industry, DNOC is made and transported as a powder often dampened with water (12% by weight) to reduce the risk of workers' exposure to dusts.

Occupational exposure is expected to occur in agriculture and in the chemical industry.

1.1.5 Kinetics and metabolism

The metabolic pathway of DNOC is qualitatively similar across several species. However, the rate of DNOC elimination varies substantially across species. Humans retain DNOC longer than other tested species.

1.1.6 Effects on laboratory mammals; in vitro test systems

1.1.6.1 Single exposure

DNOC has oral median lethal dose (LD_{50}) values ranging from 20 to 85 mg/kg body weight (b.w.) in the rat and 50–100 mg/kg b.w. in the pig. Its percutaneous LD_{50} is in the range of 600 to over 2000 mg/kg b.w. in the rat, and 1000 mg/kg b.w. in the rabbit, indicating a limited dermal absorption. Inhalation median lethal concentration (LC_{50}) values of 230 mg/m^3 for a 4-h exposure in the rat, and 40 mg/m^3 (4 h) in the cat, have been determined.

1.1.6.2 Short-term exposure

Short-term dietary administration of DNOC for up to 90 days decreased body-weight gain in rats, mice and dogs, usually without significant alteration in food consumption. At high doses the liver was affected, as shown by an increased activity of liver enzymes. Blood urea levels were also increased at high dosages.

1.1.6.3 Skin and eye irritation and skin sensitization

Application of DNOC to the skin of rabbits induced erythema and oedema, indicating an irritating effect. DNOC is a skin sensitizer in the guinea-pig and corrosive to the eyes of the rabbit.

1.1.6.4 Long-term exposure

In a long-term dietary feeding study in the rat, DNOC did not induce any treatment-related adverse effects at doses up to 5 mg/kg b.w. per day. Food consumption was found to be slightly higher (+6%) in the group receiving the highest dose than in the untreated control. This effect (i.e., higher food consumption without effect on the body-weight gain) is a consequence of the particular mode of action of the product.

1.1.6.5 Reproduction, embryotoxicity and teratogenicity

At high doses DNOC has a slight effect on reproduction in the form of reduction of body weight and litter size. Other reproduction parameters are not affected.

DNOC did not induce any teratogenic effects in pregnant rats receiving oral doses up to 25 mg/kg b.w. per day from gestation day 6 to day 15, inclusive. In rabbits, treated orally, the high dose of 25 mg/kg b.w. per day was maternally toxic, inducing mortality. At this dose level teratogenic effects, including microphthalmia or anophthalmia and hydrocephaly or microcephaly, were observed.

When administered to pregnant rabbits by cutaneous application during gestation, DNOC induced maternal toxicity at the high dose of 90 mg/kg b.w. per day, resulting in some embryotoxicity but not teratogenicity. No evidence of teratogenicity or embryotoxicity was recorded in mice treated orally or intraperitoneally during pregnancy.

1.1.6.6 Mutagenicity

On the basis of all the data available, the mutagenicity of DNOC remains equivocal.

EHC 220: Dinitro-ortho-cresol

1.1.6.7 Carcinogenicity

In a long-term dietary feeding study in rats, DNOC did not cause an increased incidence of any type of tumour.

1.1.7 Effects on humans

DNOC has caused acute poisoning in humans. Symptoms associated with DNOC toxicity are restlessness, a sensation of heat, flushed skin, sweating, thirst, deep and rapid respiration, tachycardia, severe increase of body temperature, and cyanosis leading to collapse, coma and death. Effects are enhanced at high environmental temperature. These effects are consistent with the proposed mechanism of action of DNOC.

1.1.8 Effects on organisms in the environment

DNOC has little effect on micro-organisms in the soil at recommended application rates. Acute toxicity to aquatic organisms is very variable, even within animal groups with LC_{50} values ranging from 0.07 to 5.7 mg/litre; fish were the most sensitive species in laboratory tests. Calculated toxicity exposure ratios (TERs) for aquatic organisms indicate some risk from spray drift. Application of a 5-m buffer zone reduces risk factors to acceptable levels.

DNOC is acutely toxic to honey bees but exposure is likely to be low; hazard quotients for honey bees indicate low risk. TER for earthworms (LC_{50} at 17 mg/kg soil) indicates moderate risk following use of DNOC as a desiccant.

The high acute toxicity of DNOC for birds and mammals is unlikely to be manifest in the environment because exposure is likely to be low. This conclusion is supported by limited reports of incidents in the field. Further characterization of risk is not possible because field information on residues and effects is not available.

1.2 Conclusions

When used according to registered recommendations, together with application of usual individual protective measures, exposure to DNOC is greatly reduced to levels that do not cause systemic toxicity.

Summary and Conclusions

Given the present use patterns of the plant protection product containing DNOC as the active ingredient, there are no detectable residues in treated crops, and thus no exposure of the general population.

DNOC is a skin sensitizer in guinea-pigs.

Agricultural use as a desiccant and on dormant fruit crops leads to calculated risk factors indicating possible adverse effects on aquatic organisms (from spray drift) and earthworms. Other organisms in the field are unlikely to be adversely affected because exposure will be low. No risk assessment was attempted for possible other uses of DNOC (such as locust control) because of lack of information on application rates and methods.

2. IDENTITY, PHYSICAL AND CHEMICAL PROPERTIES, ANALYTICAL METHODS

2.1 Chemical identity

Chemical formula: $C_7H_6N_2O_5$

Chemical structure:

[Structure of 2-methyl-4,6-dinitrophenol: benzene ring with OH, CH₃, and two NO₂ groups]

Relative molecular mass: 198.13

Common name: DNOC (ISO, WSSA, BSI, JMAF)

Chemical names: 4,6-dinitro-*ortho*-cresol (IUPAC)
2-methyl-4,6-dinitrophenol (CA)
2,4-dinitro-*ortho*-cresol
3,5-dinitro-2-hydroxytoluene
2,4-dinitro-6-methylphenol

Synonyms: DNC; ENT 154

Common trade names: Antinonin; Bonitol; Dinitrol; Technolor; Trifocide; Trifina; Veraline

Trade names no longer in use: Elgetol; Extar A; Nicyl; Nitrador; Sandoline; Selinon; Sinox

CAS registry number: 534-52-1

CIPAC number: 19

EEC number: 208 601 1

UN number: 1598

Identity, Physical and Chemical Properties, Analytical Methods

Table 1. Physical and chemical properties of DNOC

Property	Characteristics	Reference
Physical state	yellow, crystalline, solid	Jongerius & Jongeneelen (1991)
Crystal structure	triclinic	Jongerius & Jongeneelen (1991)
Purity of the technical product	97.45%	Sainsbury et al. (1995)
	95–98%	Tomlin (1997)
Molecular weight	198.13	Tomlin (1997)
Melting point	88.2–89.8 °C	Hope et al. (1995)
Boiling point	312 C	Jongerius & Jongeneelen (1991)
Vapour pressure	1.6×10^{-2} Pa at 25 °C	Howarth et al. (1995)
Relative density	1.58 at 20 °C	Hope et al. (1995)
Solubility in water (20 °C)	0.214 g/litre at pH 4 6.94 g/litre at pH 7 33.3 g/litre at pH 10	Hope et al. (1995)
Solubility in organic solvents (at 20 °C)		Hope et al. (1995), Tomlin (1997)
toluene	251 g/litre	
methanol	58.4 g/litre	
dichloromethane	503 g/l	
acetone	514 g/litre	
hexane	4.03 g/litre	
log P_{ow}	1.78 at pH 4 8.67×10^{-2} at pH 7	Hope et al. (1995)
Dissociation constant (pK_a)	4.48 at 20 °C	Hope et al. (1995)
	4.9 and pH limits 3–8.5	Heimlich & Nolte (1993)
Vapour density	6.84 (air = 1)	Jongerius & Jongeneelen (1991)
Saturation vapour concentration (20–25 °C)	0.56–1.0 mg/m^3	Jongerius & Jongeneelen (1991)
Conversion factor (at 760 mmHg and 20 °C)	1 mg/m^3 = 0.12 ppm 1 ppm = 8.24 mg/m^3	Jongerius & Jongeneelen (1991)
Flammability	no auto-ignition below 400 °C	Tremain & Bartlett (1995)
Stability in water	DT_{50} >1 year	Tomlin (1997)
Photolysis	PT_{50} 253 h (20 °C)	Tomlin (1997)

2.2 Physical and chemical properties

Some DNOC physicochemical properties are given in Table 1.

Like all other dinitrophenols, DNOC is a pseudoacid and readily forms water-soluble salts with alkalis (Metcalf, 1978; HSDB, 1994). At pH 4.4, more than 50% of the DNOC in water exists as the free anion. The concentration of DNOC in ionized form increases as the pH increases, and at pH 7 or above 100% of DNOC will be in the ionized form. Therefore, at physiological pH DNOC is either ionized or bound to macromolecules (i.e., albumin) (King & Harvey, 1953b).

2.3 Analytical methods

The analytical methods used to quantify DNOC in environmental and biological samples, particularly those approved and currently used by federal agencies and organizations, are listed in Tables 2 and 3, respectively. These tables also include some modifications to previously used methods that allow lower detection limits, and/or improve accuracy and precision.

Levels of DNOC in environmental and biological samples can be measured following several extraction or clean-up steps. These steps might include liquid–liquid extraction, solid phase extraction or solid phase microextraction. Both HPLC and GC with several detection methods are used for final separation and quantification.

All analytical methods used for measuring DNOC in biological samples listed in Table 3 rely on spectrophotometry for final quantification, with the exception of those of Hopper et al. (1992) and Diepenhorst et al. (1995). False positive results may be obtained by these methods because of abnormally high bilirubin or carotene levels in the blood (Jongerius & Jongeleenen, 1991).

Table 2. Analytical methods for measuring DNOC in environmental samples

Type of sample	Preparation	Analytical method	Detection limit	Recovery (%)	Reference
Technical and formulated products	Dissolve sample in methanol or acetone	HPLC/UV	2 ng	No data	Farrington et al. (1982); Yao et al. (1991)
Technical products	Dissolve sample in methanol	HPLC/ELCD	0.1 ng (oxidative) 0.4 ng (reductive)	No data	Yao et al. (1991)
Air	Draw air through filter and a midget bubbler in series. DNOC extracted into ethylene glycol and 2-propanol added before analysis	HPLC/UV (method S166)	0.070 mg/m (8 ppb) for 180-litre sample	104 for 0.07 mg loaded on to filter	NIOSH (1984)
Water	Sample adjusted to pH 6.1 by buffer	HPLC/AdSV HPLC/DPP	0.1 µg/litre (AdSV) 1.5 µg/litre (DPP)	No data	Benadikova & Kalvoda, (1984)
Water	Extract reconstituted in methanol-acetonitrile acetic acid (20:78.5:1.5 v/v)	HPLC/UV	No data	97	Tripathi et al. (1989)

Table 2 (contd)

Drinking-water, atmospheric water	Acidify sample, add salt, and extract continuously with methylene chloride. Dry, reduce volume, and solvent exchange to hexane. Derivatize with acetic anhydride	GC/NPD	0.20 µg/litre (0.2 ppm)	102 (5.5% RSD)	Herterich (1991)
Drinking-water, groundwater	Acidify water, add sodium sulfite, and pass through SPE cartridge of Carbopak. Elute with methanol/methylene chloride; reduce volume	HPLC/UV	0.009 µg/litre (9 ppb)	96	Di Corcia & Marchetti (1992)
Groundwater	Acidify to pH 2, saturate with salt, and extract using SPME	GC/MS	0.070 µg/litre (0.07 ppm) (5.6% RSD)	No data	Buchholz & Pawliszyn (1993)
Groundwater, sediment	Extract acidified water with methylene chloride, reduce volume and solvent exchange to 2-propanol	GC/FID (Method 8040)	160 µg/litre	$0.84C - 1.01$ where C is the true value of concentration in µg/litre	US EPA (1986a)
Groundwater, soil, solid waste	Extract acidified water with methylene chloride, reduce volume and exchange into 2-propanol. For other matrices, mix with anhydrous sodium sulfate and extract (soxhlet or sonication) with methylene chloride. Reduce volume. Clean up with silica gel if needed	GC/MS (Method 8270)	50 µg/litre (50 ppm water); 3.3 mg/kg (ppm soil/sediment)	$1.04C - 28.04$ where C is the true value of concentration in µg/litre	US EPA (1986b)

Table 2 (contd)

Matrix	Procedure	Technique	Detection limit	Recovery	Reference
Waste water	Extract acidified sample with methylene chloride; concentrate and exchange solvent to 2-propanol	GC/FID (Method 604)	16 µg/litre (16 ppm)	83 at 100 µg/litre	US EPA (1984a)
Waste water	Extract acidified sample with methylene chloride; concentrate	GC/MS (Method 625)	24 µg/litre (24 ppm)	93 at 100 µg/litre	US EPA (1984b)
Waste water	Extract acidified sample with methylene chloride, dry and reduce volume. Add deuterated standards	GC/MS isotope dilution (Method 1625)	20 µg/litre (20 ppm)	77–133 at 100 µg/litre	US EPA (1984c)
Rain and snow	Extract acidified sample with methylene chloride; concentrate	HPLC/PDD	No data	No data	Alber et al. (1989)
Soil	Extract with methylene chloride; evaporate to dryness and dissolve residue in alkaline methanol/water	HPLC/UV	0.005 mg/kg (5 ppb)	85–105	Roseboom et al. (1981)
Soil	Soxhlet extraction of clay loam using hexane : acetone (1 : 1). Reduce volume	GC/MS	No data	63.4 at 6 mg/kg	Lopez-Avila et al. (1993)
Various crops	Extract macerated or homogenized sample with methylene chloride; evaporate to dryness and dissolve in potassium carbonate/methanol mixture	HPLC/UV	0.005 mg/kg (5 ppb)	82–105 at 0.05 mg/kg %RSD range 4–13%	Roseboom et al. (1981)

Table 2 (contd)

Various crops	Homogenize sample in blender, adding distilled water as needed. Add Florisil to form free flowing mixture and pack into a column with a sodium sulfate layer at bottom. Elute with methylene chloride : acetone (1 : 1) or ethyl acetate. Reduce volume	GC/ELCD	0.001 mg/kg (1 ppb)	69–79 at 0.01– 0.5 mg/kg	Kadenczki et al. (1992)
Fatty and non-fat foods	Mix fatty sample with methanol, sulfuric acid and potassium oxalate and, non-fat samples with sulfuric acid and methanol; extract both with petroleum ether or methylene chloride; clean-up by gel permeation chromatography, methylate and clean up with Florisil	GC/NPD	No data	45–50 (fatty foods) >80 (non-fat foods)	Hopper et al. (1992)

ᵃThere are absolute detection limits.

AdSV, adsorptive stripping voltametric detector; DPP, differential pulse polarographic detector; ELCD, electrochemical detector; FID, flame ionization detection; GC, gas chromatography; HPLC, high-performance liquid chromatography; HRGC, high-resolution gas chromatography; MS, mass spectometry; NPD, nitrogen phosphorus detector; PDD, photodioue array detector; RSD, relative standard deviation; SPE, solid phase extraction; SPME, solid phase microextraction; UV, ultraviolet detector; v/v, volume per volume.

Table 3. Analytical methods for measuring DNOC in biological samples

Sample matrix	Preparation method	Analytical method	Sample detection limit	Recovery (%)	Reference
Animal tissue	Extract sample mixed with methanol, sulfuric acid, and potassium oxalate with petroleum ether; clean up by gel permeation chromatography, methylate, and clean up with Florisil	GC-NPD	No data	45–50	Hopper et al. (1992)
Urine, kidney, liver, brain (DNOC and metabolite 4-amino-2-methyl-6-nitrophenol)	Hydrolyse sample directly or after acetone extraction; extract with petroleum ether	Spectrophotometric	No data	No data	Truhaut & de Lavaur (1967)
Serum	Dilute with water; add sodium chloride and sodium carbonate and extract with methyl ethyl ketone	Spectrophotometric	<0.5 mg/litre	No data	Parker (1949)
Serum	Samples were acid coagulated then serum separated by centrifugation	HPLC/UV	0.05 µg/g	91.0	Diepenhorst et al. (1995)

Table 3 (contd)

Tissue	Dilute homogenized tissue with water; add sodium chloride and sodium carbonate; extract with methyl ethyl ketone	Spectrophoto-metric	No data	Parker (1949)
Urine (DNOC and metabolite 4-amino-2-methyl-6-nitrophenol)	Acidify and subject to continuous extraction with diethyl ether	Spectrophoto-metric	No data	Smith et al. (1953)
Urine	Add sodium chloride and sodium carbonate; extract with methyl ethyl ketone	Spectrophoto-metric	No data	Parker (1949)

GC, gas chromatography; NPD, nitrogen phosphorus detection device.

3. SOURCES OF HUMAN AND ENVIRONMENTAL EXPOSURE

3.1 Natural occurrence

DNOC does not occur naturally.

3.2 Anthropogenic sources

3.2.1 Uses

DNOC was first introduced as an insecticide in 1892 and as a herbicide in 1932 (Gasiewicz, 1991). It is registered in a number of countries for use as an acaricide, larvicide and ovicide to control the dormant forms of many insects in orchards. It is applied during the winter on pome and stone fruits and grapes ("winterwash"). Registered DNOC uses specify rates ranging from 840 to 8400 g/ha of active ingredient. One spray application is made in the dormant period of deciduous crops.

DNOC is also used as a desiccant in potatoes. It is sprayed once or twice on seed potatoes between July and September to desiccate the haulms in order to prevent virus and disease contamination of the tubers, and incidentally to facilitate mechanical harvesting. The registered rates of application of DNOC range from 2.5 to 5.6 kg/ha.

DNOC is formulated as emulsifiable concentrate (EC) for use as a potato haulm desiccant and as a suspension concentrate (SC) for winter treatment on fruit trees. Other types of formulation include pastes (PA) and oil-miscible liquids (OL). It is understood that DNOC is still used as a desiccant for crop potatoes and in locust control in developing countries. However, details of sources, application rates and methods are not available.

Although the use of DNOC as a pesticide has currently declined, and also because it has been banned in some countries (see for instance EC, 1999), there are still significant volumes of obsolete stocks of this chemical around the world, especially in developing countries. The German Agency for Technical Cooperation (GTZ) has helped in disposing of 57.6 tonnes of DNOC in the United Republic of Tanzania by incineration in a cement kiln (GTZ, 1997).

More than 14 tonnes of obsolete DNOC have been located in Zambia (Wodageneh, 1997).

The main current use of DNOC is in the plastics industry as an inhibitor of polymerization in styrene and vinyl aromatic compounds. It is also used as an intermediate for synthesis of other fungicides, dyes and pharmaceuticals (Hawley, 1981; US EPA, 1988).

3.2.2 Worldwide sales

The worldwide annual production of DNOC was approximately 2000 tonnes in the 1950s, all of which was used in agriculture. Currently, of the 600 tonnes or so manufactured annually, 400–500 tonnes are used in industry and 100–200 tonnes as an agrochemical.

4. ENVIRONMENTAL TRANSPORT, DISTRIBUTION AND TRANSFORMATION

4.1 Transport and distribution between media

4.1.1 Air

DNOC has a vapour pressure of 1.6×10^{-2} Pa at 25 °C and a solubility in water of 6.94 g/litre at pH 7 and 20 °C. As a result, it has a Henry's law constant of 2.46×10^{-7} atm · m³/mol. On this basis, DNOC has no potential to volatilize from surface waters.

4.1.2 Water

DNOC is only moderately adsorbed on aquatic sediments (Vonk & van der Hoven, 1981).

4.1.3 Soil

Adsorption studies with DNOC are complicated by its changing dissociation with pH and its rapid degradation in soil. Jafvert (1990) studied the adsorption of DNOC to 13 well-characterized soils from the midwestern United States. Adsorption was monitored after 24 h, avoiding degradation of the compound. For 11 soils where the pH was in the range 7.0–8.3, the adsorption coefficient correlated positively with organic carbon content. DNOC was adsorbed very strongly to a sediment with very low organic carbon content at pH 4.47, reflecting the behaviour of the undissociated compound at this acidic pH. Values of K_p (the adsorption coefficient) ranged from -0.16 to 5.93 for DNOC in the 13 soils; the negative value for one sediment reflects repulsion of the organic anion in this case. Using one of the soils, distribution ratios were determined as a function of pH and the overall adsorption modelled; over the pH range 4–9.5, the fraction in the aqueous phase increased from approximately 20% to more than 90%. In a study using three soil types under OECD guideline 106 (Jonas, 1995), percentage adsorption after 16 h was less than 16% (range 13.6–15.4%). Continuation of the test showed that no plateau was reached, reflecting degradation of the compound. Results at 3 days showed adsorption of up to 79% in one soil at a concentration of 0.024 mg/litre DNOC; however, adsorption was concentration dependent with only 22% adsorption in the same soil at 3 mg/litre DNOC.

On the basis of the Gustafson (1989) groundwater ubiquity score, DNOC is considered to have a limited potential to leach from soil to groundwater.

4.2 Degradation

The half-life of DNOC in surface water ranges from 3 to 5 weeks (Vonk & van der Hoven, 1981).

DNOC is metabolized in soil. One bacterium of the *Arthrobacter* species is capable of using the compound as its source of carbon and nitrogen (Gasiewicz, 1991). It was also demonstrated that DNOC is rapidly inactivated in soil by a form of *Corynebacterium simplex* with formation of nitrite (Jensen & Gundersen, 1955). The biological decomposition of DNOC in soils was reviewed by Jensen (1966).

Tewfik & Evans (1966) have isolated a *Pseudomonas* species able to degrade DNOC in soils. The degradation of DNOC by 31 strains of *Rhizobium* and 5 strains of *Azotobacter* has been described (Hamdi & Tewfik, 1970); this microflora is important in nitrogen fixation.

The degradation of DNOC in three types of standard soils was investigated over a period of 88 days, at 20 °C, in the dark, at an application rate of 4.9 mg ^{14}C-labelled DNOC/kg (dry weight) of soil. This is equivalent to a field application rate of 5 kg DNOC/ha. The DT_{50} was determined to be 1.7, 5.9 or 12 days, depending on the soil type. The main final degradation product of the aromatic ring was carbon dioxide, representing 39% of the applied radioactive dose; the main non-volatile metabolite was 2-methyl-4-nitrophenol, representing 40% of the applied radiocarbon between day 10 and day 20, and declining thereafter. The amount of bound residues in soil after extraction with organic solvents increased over the course of the study to reach 37% (Bieber, 1995). The presence of 2-methyl-4-nitrophenol as a decomposition product of DNOC in soil was confirmed by Verheij & van der Graaf (1995) by combined liquid chromatography–mass spectrometry (LC–MS) and tandem mass spectrometry (MS–MS).

4.3 Crop uptake

DNOC is not a systemic compound in plants. One of its main uses is as a winterwash on fruit trees to destroy the dormant forms of eggs and larvae of various insects. It is sprayed on to the trees during the winter at a time when there are no leaves to absorb the product. It is also used as a desiccant of the haulms of seed potatoes. The haulms are quickly killed and dried, thus preventing any absorption and translocation of the product to the potato tubers. No residues are found at harvest in fruits from trees treated during the winter or in potatoes treated for haulm killing.

5. ENVIRONMENTAL LEVELS AND HUMAN EXPOSURE

5.1 Environmental levels

5.1.1 Air

Cresols, including DNOC, have been detected in atmospheric air in the ng/m^3 range as well as in different condensed phases in the atmosphere (rain, fog and snow) up to about 100 µg/litre (Tremp et al., 1993).

DNOC has been identified in extracts of rain (Leuenberger et al., 1988; Alber et al., 1989), and snow (Alber et al., 1989). One pathway through which DNOC can enter the atmosphere is from overspray during use on agricultural products. DNOC has been detected in rain throughout the year, and its concentrations in rain did not show a trend with seasonal applications to crops (Leuenberger et al., 1988). These observations, and its low volatility, indicate that DNOC most likely enters the atmosphere through another mechanism. The low air–water partition coefficient of DNOC allows it to be scavenged effectively by precipitation, and enriched in humid aerosols, fog, clouds and rain droplets.

5.1.2 Water and soil

DNOC has been detected in five groundwater samples at a maximum concentration of 35 µg/litre in California, where DNOC had been used as a pesticide (Hallberg, 1989). It was also occasionally detected in groundwater, ponds and streams in cultivated areas in Denmark over a 2-year survey period: for example, 1 finding out of 38 samples of groundwater at 0.05 µg/litre, and 2 out of 9 samples in pond water at a concentration range of 0.12–0.18 µg/litre (Morgensen & Spliid, 1995). It was also detected in the Klang river in Malaysia at concentrations ranging from 3.2 to 78.8 µg/litre (Tan & Chong, 1993).

Biodegradation is the most significant process for removal from water and soil.

5.1.3 Food and feed

See section 5.2.

5.2 General population exposure

5.2.1 Oral exposure

Although the potential for exposure to the general population exists through the ingestion of treated foods, the normal use patterns of DNOC do not allow for penetration into plant tissues. Residue levels of DNOC have not been detected in treated fruits and potatoes. Bioaccumulation into aquatic or terrestrial wildlife is not expected because of rapid biodegradation. DNOC has been detected in both ground and surface waters, but not in drinking-water. The compound was reviewed by the FAO/WHO Joint Meeting on Pesticide Residues (JMPR) in 1963 and 1965 (FAO/WHO, 1964, 1965), and no acceptable daily intake (ADI) or maximum residue limit (MRL) was established.

5.2.2 Inhalation exposure

DNOC can enter the atmosphere from spraying during agricultural use and subsequent dispersion from treated surfaces. Concentrations in ambient air have been detected in the ng/m^3 range and in rain, fog and snow at levels up to 100 µg/litre (Tremp et al., 1993). At these concentrations, no significant exposure to the general population is expected.

5.3 Occupational exposure during manufacturing, formulation and use

Levels of exposure of agricultural workers during normal field use (seed potato haulm desiccation) were measured (Heuts, 1993). During the spraying season, 11 male agricultural workers (independent farmers and professional sprayers) were monitored while applying a DNOC formulation containing 200 g/litre of active ingredient. The total exposure time during the season ranged from 2 to 22 h per worker. In only 2 cases was DNOC detected in the blood of independent farmers, at levels of 0.6 and 0.8 mg/litre of blood. DNOC was not detected in the blood of professional sprayers (the sensitivity of the colorimetric analytical method used was 0.5 mg/litre).

In the plastics industry, workers may be exposed to dusts when the damping water is removed before use. DNOC is used to inhibit immediate polymerization of styrene during the distillation and purification stages of manufacture. During the process, DNOC remains in the distillation columns, thereby ensuring that the finished styrene monomer contains no residues. The distillation process allows recycling of some DNOC, and the remaining DNOC-rich by-products are incinerated, thereby greatly reducing the risk of occupational and environmental exposures.

Table 4 summarizes the time-weighted average (TWA) values for occupational exposures.

In the former USSR, a maximum allowable concentration in the working zone (MAC_{WZ}) of 0.05 mg/m^3 as a mixture of vapour and aerosol; a value of 0.002 mg/m^3 for the lightest short, single exposure, tentatively safe exposure level in the atmosphere of residential areas ($TSEL_{hm}$); and a value of 0.05 mg/litre for surface water were established (Izmerov et al., 1982).

WHO (1982) indicates: "there exists a fair agreement, although no adequately valid relationship has yet been established, that − on the basis of *human* data − a blood DNOC level below 20 mg/litre will probably not lead to manifest health impairment"; the Dutch Expert Committee on Occupational Standards recommended a biological operator exposure limit (BOEL) in whole blood of 10 µg/ml. In their report, prepared on behalf of the Industrial Medicine and Hygiene Unit of the Health and Safety Directorate of the Commission of the European Communities, Jongerius & Jongeneelen (1991) recommended, based on human exposure data, a BOEL of 10 µg/ml in serum or 5 µg/ml in whole blood for workers not exposed to heat stress.

Table 4. TWA values for DNOC occupational exposures

Country	TWA (mg/m³ per 8 h)	Year established
Argentina	0.2	1991
Canada	0.2	1994
Finland	0.2	1996
Denmark	0.2	1996
Germany	0.2	1996
Mexico	0.2	1991
Netherlands	0.2	1996
Norway	0.2	1996
UK	0.2	1996
USA (OSHA)	0.2	1996
USA (NIOSH)	0.2 (10 h)	1996

Source: UNEP Chemicals (IRPTC) (1999).

6. KINETICS AND METABOLISM

DNOC may be absorbed through the skin as well as by ingestion or inhalation of aerosols. The skin is the principal route of exposure in agricultural workers. The metabolic pathway of DNOC is identical in several non-ruminant mammalian species, but the rate at which it is cleared from the organism varies between species. In ruminants, DNOC undergoes an initial phase of bacterial metabolism in the rumen before it is absorbed into the blood.

6.1 Absorption

Formulated DNOC was applied to the shaved skin of male and female rats in a single dose of 18.1 mg DNOC/kg b.w. and was kept in contact with the skin for 8 h (Fabreguettes, 1993). Two formulations were used: water-based (621 g a.i./litre) and oil-based (130 g a.i./litre). Blood samples were taken after 15 and 30 min, and 1, 2, 4, 24, 48, 72 and 96 h of contact. The results showed that DNOC in aqueous formulation was absorbed through the rat's skin at a limited rate: at the peak level the plasma concentration represented only 2.5% (14–17 µg/ml of blood) of the applied dose. The peak plasma level occurred at 24 h in female rats, and at 48 h in male rats. The average $t_{1/2}$ absorption time (time from exposure required to reach 50% of DNOC blood peak concentration) was 15 h for males and 13 h for females. The average $t_{1/2}$ elimination time (time required for DNOC blood concentration to decrease to 50% of its peak value after reaching a peak) was 24 h. After 96 h, less than 1% of the applied dose remained in the plasma, and approximately 1% in the skin.

When applied under the same conditions as an oily formulation, the peak plasma concentrations represent 5.0% of the applied dose in males and 5.8% in females (38–45 µg/ml of blood). The peak plasma level occurred after 8 h in males and 24 h in females. The average $t_{1/2}$ absorption was 2.8 h and the average $t_{1/2}$ elimination was 34 h.

DNOC is more readily absorbed through the skin in oily formulation than when in aqueous solution; the peak plasma concentration is higher and is reached earlier. However, elimination remains fairly rapid, and the residual plasma and skin levels are comparable for the two types of formulation (Fabreguettes, 1993).

Following a single dose by gavage, the maximum plasma concentration is reached in 2–4 h in rats, and in 4–6 h in rabbits (Gasiewicz, 1991).

6.2 Distribution and accumulation

Experimental work has shown that the concentration of DNOC in the blood is much higher than in any other tissues. Over 90% of the DNOC in blood is found in the plasma (Parker et al., 1951). Orally administered DNOC does not increase in the blood of laboratory animals to the same extent as in humans, possibly because of a faster elimination rate in these animals. Blood levels in rats, dogs and rabbits did not increase significantly after a second oral dose was administered (Parker et al., 1951; King & Harvey, 1953a), whereas, in humans, daily oral administration of approximately 1 mg/kg b.w. for 7 consecutive days induced a continuing increase in plasma levels. In the rat, 8 daily doses of 5 mg/kg b.w. produced an average blood level of 12 µg/ml. In humans, five daily doses of approximately 1 mg/kg b.w. resulted in blood levels of 15–20 µg/ml (Harvey et al., 1951; King & Harvey, 1953b).

A single oral dose of ^{14}C-labelled DNOC (0.4 mg/kg b.w.) given to two rats resulted in the following tissue distribution:

- In one rat 24 h after dosing, 15% of the administered dose was found in the blood, 6.6% in the gastrointestinal tract, 5% in the liver, 0.08% in the spleen, 0.94% in the kidneys and 28% in the residual carcass. The faeces contained 10.1% of the radioactivity and urine 28.7% with a total recovery at 24 h of 94.4%.

- In the other rat, 72 h after dosing, these percentages had been approximately halved, except for a lower percentage in blood (5.5%), a similar percentage in kidneys (0.9%) and a total recovery of 98.8% (Leegwater et al., 1982).

Harvey et al. (1951) reported blood DNOC concentrations in volunteers given capsules containing 75 mg DNOC every day. In 3 subjects capsules were given for 5 days and blood DNOC, measured 4 h after dosing, rose steadily to about 20 µg/g, except in one subject in whom DNOC in blood rose on the fifth day to about 40 µg/g. He received the largest dose (1.27 mg/kg b.w.). In two other subjects capsules were given for 7 days. In one subject blood DNOC rose to about 38 µg/g whereas in the other blood DNOC peaked after the sixth and seventh capsules to 40 and 48 µg/g, respectively. In the latter volunteer symptoms consistent with DNOC poisoning were observed.

A field study of 18 sprayers showed that a daily exposure to DNOC leads to continuous elevation of DNOC level in the blood. The plasma levels increased daily and, at the end of the season, plasma levels ranged from 11 to 88 µg/ml (van Noort, 1960).

6.3 Biotransformation

In the rabbit, 5% of DNOC single oral doses of 20 or 30 mg/kg is excreted unchanged and 1% as conjugated DNOC in 2-day urine collection. The main metabolic pathway is the reduction of DNOC to 6-amino-4-nitro-o-cresol (6-ANOC), and to a lesser extent to 4-amino-6-nitro-o-cresol (4-ANOC). Urinary content of 6-ANOC accounted for 11–12% of the administered dose. Small amounts of other metabolites, such as 4-ANOC conjugates and 3-amino-5-nitrosalicylic acid (3-ANSA), which is produced via an oxidative pathway, are also excreted in the urine (Smith et al., 1953).

Leegwater et al. (1982) and van der Greef & Leegwater (1983) have identified similar metabolites as well as two other new ones, not previously described: 4,6-diacetamido-o-cresol (DAcAOC) and 4,6-dinitro-2-hydroxymethylphenol (DNHMP) in the urine of rats treated with a single oral dose of 0.4 or 6.0 mg DNOC/kg b.w., and in the urine of a rabbit administered orally a single dose of 20 mg DNOC/kg b.w.

Based on these observations, it may be concluded that rats and rabbits metabolize DNOC along the same pathway (Fig. 1) as suggested by Leegwater et al. (1982) with slight modification.

Fig.1. Biotransformation of DNOC in the rat and rabbit

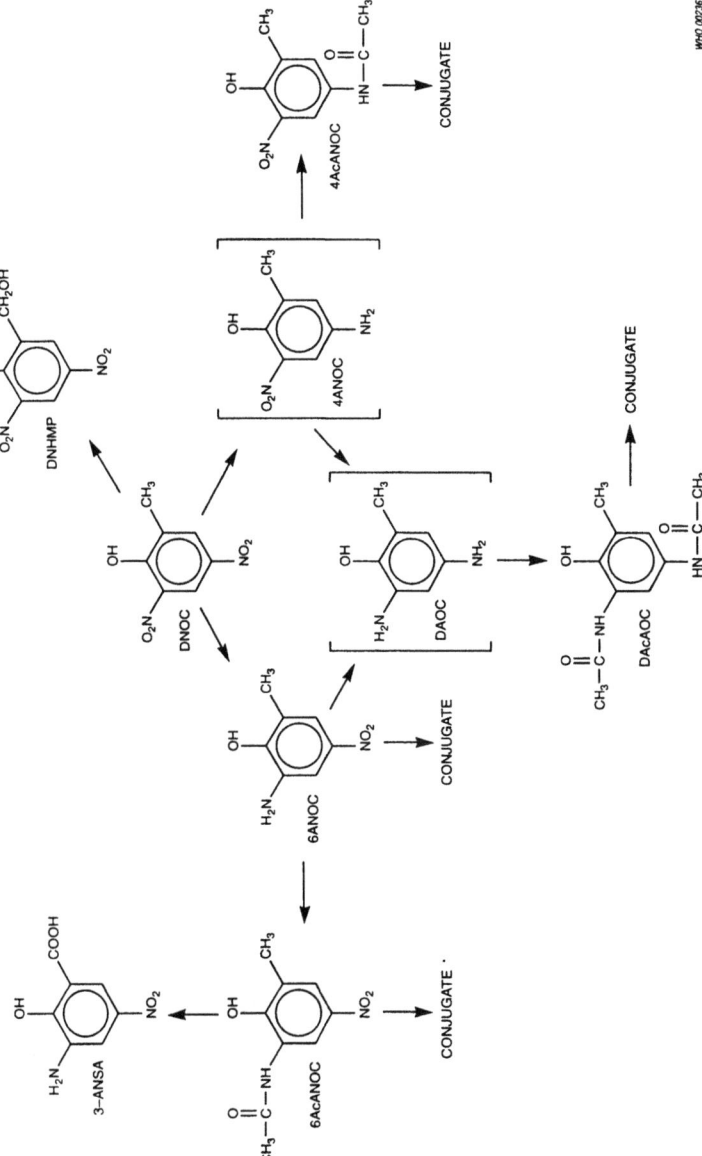

In an *in vitro* study in which DNOC was incubated with the contents of rat caecum, a rapid reduction to 6-ANOC occurred and 6-ANOC was then converted to DAOC. After 12 h of contact, 90% of the initial concentration of DNOC had been metabolized to DAOC (Ingebritsen & Froslie, 1980).

In ruminants (cattle), DNOC induces methaemoglobinaemia when administered intra-ruminally. This effect is related to the reduction process mediated by microflora that occurs in the rumen, leading to the formation of aminophenols and diaminophenols, which are known to be methaemoglobin-forming compounds (Harvey, 1958; Froslie & Karlog, 1970; Froslie 1973). The role of the microflora in the metabolism of DNOC by ruminants was confirmed experimentally in sheep by Jegatheeswaran & Harvey (1970).

6.4 Elimination and excretion

DNOC is excreted in the urine as free DNOC and acetylated conjugate 6-ANOC, conjugated as 6-acetamido-4-nitro-*o*-cresol (6-AcANOC) (WHO, 1982). In rats given a single oral dose of 0.4 mg ^{14}C-DNOC/kg b.w., the elimination half-life was determined to be 1–1.5 days (Leegwater et al., 1982). This observation is in agreement with that of King & Harvey (1953a) who determined a half-life of 26.8–28.5 h in female rats treated with either nine daily doses of 20 mg/kg b.w. or a single dose of 30 mg/kg b.w.

In female rabbits, the half-life was determined to be approximately 6.5 h. After repeated dosing in humans, the DNOC level in blood increases more than that in laboratory animals (Harvey et al., 1951), because it is excreted at a slower rate in humans than in animals (Parker et al., 1951; Pollard & Filbee, 1951). In humans, the half-life of DNOC has been calculated from blood levels measured under circumstances related to heavy occupational exposure. The half-lives so determined varied from 96 h (van Noort, 1960) to 148 h (Jastroch et al., 1978) or 153.6 h (Pollard & Filbee, 1951) in severely poisoned sprayers. Lawford et al. (1954) have demonstrated that the elimination rate in descending order was:

mouse > rabbit > guinea-pig > rat = monkey > humans.

6.5 Reaction with body components

In male guinea-pigs given intraperitoneal daily doses of one-third the LD_{50} for 30 days, six times a week, a statistically significant increase in amino sugars and sialic acid levels was observed in the serum and liver. A decrease of glycoprotein content in the albumin and α_2-globulin fractions was noted, together with an increase of glycoprotein content of the serum α_1- and γ-globulin fractions. These results suggest that DNOC increases the biosynthesis of the sugar moiety of glycoproteins by increasing the glycolysis rate and by disturbing the stability of the lysosomal membranes (Kreczko et al., 1974). Parker (1952) has shown that DNOC was reduced *in vivo* by rat-liver homogenates to 4-amino-2-nitrophenol and that 2-amino-4-nitrophenol was also generated to a lesser extent.

Van den Berg et al. (1991) found that DNOC is an *in vitro* competitor for the thyroxine (T_4) binding site on the plasma protein transthyretin. This plasma protein is a carrier for vitamin A and hormones, including T_4. Speculations suggest that DNOC may alter thyroid hormone levels in plasma, thereby affecting thyroid functions.

7. EFFECTS ON LABORATORY MAMMALS; *IN VITRO* TEST SYSTEMS

7.1 Single exposure

The acute toxicity of DNOC in several laboratory species is summarized in Table 5. Signs of acute toxicity include hyperactivity, laboured breathing, asphyxial convulsions, coma and death (NIOSH, 1978).

7.1.1 Oral exposure

Acute oral exposures to DNOC resulting in toxicity and death have been reported in rats, mice, cats and pigs at relatively similar doses (Table 5). Increases in environmental temperature enhanced the acute toxicity among rats orally dosed with DNOC. After receiving a single dose of DNOC of 20 mg/kg b.w. at 37–40 °C, 6 of 12 rats died, whereas only 2 of 12 rats died at receiving a single dose of 40 mg/kg b.w. at 20–30 °C (King & Harvey, 1953a).

7.1.2 Inhalation exposure

Rats exposed to air concentrations of 100 mg/m^3 for 4 h at an ambient temperature of 28–30 °C survived the exposure period but showed increased body temperature and respiration rates (King & Harvey, 1953b). A single exposure of 12 cats to the same air concentration of 100 mg/m^3 for 4 h resulted in the death of 4 cats (1 on day 4, 2 on day 6 and 1 on day 11), whereas a concentration of 60 mg/m^3 for 4 h induced no deaths. However, a concentration of 40 mg/m^3 killed 1 of 3 cats (Burkatskaya, 1965a). It should be noted that the experimental protocols, as well as the technologies applied in these experiments, were inadequate to determine reliably the actual concentrations to which the animals were exposed.

7.1.3 Skin exposure

When kept in contact with the intact skin of three New Zealand White rabbits for 4 h under a semi-occluding bandage, a dose of 0.5 g of technical (97.5% pure) DNOC/rabbit in 0.5 ml of distilled water spread on an area of 2.5 × 2.5 cm^2 of skin induced erythema, slight oedema and crust formation, indicative of an irritating effect (Driscoll, 1995c). The subcutaneous LD_{50} for rats decreases at higher temperatures (Jongerius & Jongeneelen, 1991). LD_{50} values

Table 5. Acute toxicity of DNOC in laboratory animals

Route	Species	LD_{50}/LC_{50} (mg/kg b.w.)[a]	Reference
Oral	rat	20 (at 37–40 °C)	King & Harvey (1953a)
Oral	rat	25	Ben Dyke et al (1970)
Oral	rat	30 (minimum lethal dose)	Ambrose (1942)
Oral	rat	31	Driscoll (1995 a)
Oral	rat	50	Spencer et al (1948)
Oral	rat	85	Burkatskaya (1965b)
Oral	cat	50	Jongerius & Jongeneelen (1991)
Oral	mouse	16	Jongerius & Jongeneelen (1991)
Oral	mouse	47	Jongerius & Jongeneelen (1991)
Oral	pig	50–100	McGirr & Papworth (1953)
Dermal	rat	200–600	Ben Dyke et al (1970)
Dermal	rat	>2000	Driscoll (1995b)
Dermal	rabbit	500 (no effect)	Burkatskaya (1965b)
Dermal	rabbit	1000	Burkatskaya (1965b)
Dermal	mouse	187	Arustamyn (1972)
Dermal	guinea-pig	200 (no effect)	Jongerius & Jongeneelen (1991)
Dermal	guinea-pig	500 (LD_{100})	Spencer et al (1948)
Inhalation	rat	100 mg/m^3 (4 h) (no effect)	King & Harvey (1953)
Inhalation	rat	230 mg/m^3 (4 h)	Dey-Hazra & Heisler (1981)
Inhalation	cat	40 mg/m^3 (4 h)	Burkatskaya (1965a)
Intra-peritoneal	rat	29.	Gasiewicz (1991)
Intra-peritoneal	mouse	24–26	Gasiewicz (1991)
Intra-peritoneal	rabbit	24	Jongerius & Jongeneelen (1991)
Intra-peritoneal	guinea-pig	23	Jongerius & Jongeneelen (1991)

of 27.7 mg/kg and 19.2 mg/kg were reported at 5–10 °C, 18–20 °C and 36–37 °C, respectively.

7.1.4 Skin sensitization

Technical grade (97.5% pure) DNOC was found to induce dermal sensitization in the guinea-pig's skin when tested according to the method of Magnusson and Kligman (Driscoll, 1995e).

7.2 Short-term exposure

7.2.1 Oral administration

7.2.1.1 Rat

In a 6-week range-finding study by Broadmeadow (1988), 5 groups of Charles River rats of both sexes (5 males and 5 females) were fed daily diets containing either 0 (control), 5, 13, 32, 80 or 200 mg of 99.5% pure DNOC/kg of feed for 6 consecutive weeks (equivalent to 0 (control) 0.44, 1.17, 2.89, 7.24 and 18.6 mg/kg b.w., respectively). The animals were observed twice a day for behavioural changes, clinical signs and mortality. Food consumption was measured 3 times a week and a weekly average calculated. Individual body weights were measured 3 times a week. Haematological, biochemical and urinary parameters were evaluated during the last week of treatment. At the end of the 6-week treatment period, all the animals were killed and necropsied.

No treatment-related mortality was recorded. No significant effects were observed on food consumption and haematology. The body-weight gain of the females of the two highest concentrations was significantly lower ($p < 0.001$) than that of the controls, whereas the food conversion factor in these two groups was slightly elevated. This effect is directly related to the particular mode of action of the compound. A slight, but statistically significant ($p < 0.05$), decrease in alanine aminotransferase (ALT) activity was observed in the animals treated with 80 and 200 mg/kg DNOC in the diet. The blood urea level was slightly, but significantly, elevated ($p < 0.05$) in females of the two high-dose groups. Body temperature was not affected. No abnormalities were seen at necropsy and consequently no histopathological examination was performed (Broadmeadow, 1988).

A 90-day feeding study was performed in Wistar rats. Groups of 10 males and 10 females were administered diets containing either 0 (controls), 10, 100, 200 or 400 mg DNOC of unspecified purity/kg in their diet (equivalent to 0, 2.5, 5.0, 10.0 or 20 mg/kg b.w. per day) for 90 consecutive days. Although not adequately reported, the following results were observed. At the highest feed concentration (400 mg/kg diet), 25% of the rats died during the course of the study. Mortality was also observed in the next 2 doses (2 rats at 200 mg/kg diet and 1 at 100 mg/kg diet). Body-weight gain was severely depressed at the high dose and also at the next lower dose, and to a lesser extent at the next lower dose of 100 mg/kg diet. Food consumption was depressed in the high-dose group and slightly in the next lower group of 200 mg/kg diet. There was a sharp increase in ALT activity in the high-dose group, but only one male and one female in this group were evaluated. Increased levels of glucose and urea were recorded in the two highest treatment concentrations in both sexes. Urea level was also increased in the 100 mg/kg diet group, but in male rats only. T_3 and T_4 levels were decreased at all dose levels. Pyruvate was decreased in a dose-related manner. No effects were seen on the lactate blood level or the organ weights. Some histopathological alterations were observed in the high-dose animals only: fewer acidophilic cells in the pituitary, atrophic islets of Langerhans, no corpora lutea in the ovaries, lower spermatogenesis and atrophic thymus (den Tonkelaar et al., 1983).

From these results it is obvious that the two highest concentrations of 400 and 200 mg/kg diet (equivalent to 20 and 10 mg/kg b.w. per day) were well above the maximum tolerated dose (MTD), and that the alterations observed in the animals treated at these levels were most likely related to the poor physiological and nutritional state of the sick animals, as indicated, in particular, by the high mortality rate and the severe body-weight depression in the high-dose group of animals. Some limited effects were seen in animals exposed to 100 mg/kg diet (equivalent to 5 mg/kg b.w. per day): decrease in body weight, increased blood urea level in males only, and some non-dose-related decrease in T3 and T4 levels. The lowest concentration of 10 mg/kg diet (equivalent to 2.5 mg/kg b.w. per day) may be considered as the no effect level (NOEL).

In a study by Spencer, concentrations up to and including 0.01% (100 mg/kg diet) did not affect the body-weight gain of the rats. At 0.02% (200 mg/kg diet) the body-weight gain was 7–9% below that of the controls. At 500 mg/kg diet the growth of the animals was well below that of the other groups. No effect was observed on the organ weights. Blood urea levels were increased when compared to controls. No histopathological alterations were recorded in the lung, heart, kidney, liver, adrenal, pancreas, testes and stomach. There was a depletion of the body fat (Spencer et al., 1948).

7.2.1.2 Mouse

A 13-week range-finding feeding study was performed in mice (18 mice/group) fed a diet containing DNOC in concentrations equivalent to a daily intake of 0, 1.0, 5.0 and 10.0 mg DNOC/kg. No clinical signs associated with an effect of the compound were observed. No mortality was recorded. The body-weight gain, as well as the food consumption, were comparable across all three groups. The total T_4 level was markedly decreased in the medium- and high-dose groups of male mice only. A treatment-related moderate decrease in T_3 blood level was seen in female mice. No treatment-related histopathological alterations were observed (Kelly, 1995).

7.2.1.3 Dog

For 90 consecutive days 16 male and 16 female beagle dogs were fed diets containing 0 (controls), 4.0, 20.0 or 100 mg of DNOC (purity 99.8%) per kg diet. These dietary concentrations were equivalent to a daily intake of, respectively, 0, 0.17, 0.89 and 4.82 mg DNOC/kg b.w.

No mortality was recorded during the course of the study, nor was any sign of adverse effect noted. A poorly characterized hyperactivity was reported in an unspecified number of dogs of the mid- and high-dose groups during the last few weeks of the study (onset not specified). Food consumption was not changed at any dose level, and, although there is a reduction in body-weight gain in males and females, it is more pronounced in males. This effect is likely to be related to the mechanism of action of DNOC. The prothrombin time was slightly reduced in all groups at the 49-day dosing interval and, in the two highest concentrations, at day 85 of treatment. However, this observation has little importance in view of the wide variations observed in the reference values from the untreated controls, as well as within the groups. Some biochemical parameters showed some isolated, or not dose-related, variations during the course of the study. Kidney and liver functional tests did not show any alterations. No histopathological modifications attributable to the treatment were observed (Til, 1980).

Effects on Laboratory Mammals: In Vitro *Test Systems*

7.2.2 Inhalation

7.2.2.1 Cat

Burkatskaya (1965a) reported that death occurred in 2 out of 3 cats exposed to 2.0 mg DNOC/m^3 for 4 h/day during a period of 30 days. Three cats exposed to 0.2 mg DNOC/m^3 for 4 h/day for either 60 or 90 days did not show any severe adverse effects.

7.3 Skin and eye irritation; skin sensitization

A volume of 0.1 ml of technical (97.5% pure) DNOC placed into the conjunctival sac of three rabbits was corrosive to the eyes (Driscoll, 1995d). No subacute repeated dose studies are available.

DNOC has been demonstrated to be a skin sensitizer in the guinea-pig (Driscoll, 1995e).

7.4 Long-term exposure

7.4.1 Rat

Groups of 50 male and 50 female F-344 rats were exposed to diets containing concentrations of 0 (controls), 2.5, 15 or 100 mg of 99.5% pure DNOC/kg diet for 104 consecutive weeks. These concentrations were equivalent to a daily intake of 0.12, 0.75 and 5.03 mg/kg b.w. in females, and to 0.10, 0.59 and 4.12 mg/kg b.w. in males. Ten males and 10 females were added to the 2 lowest concentration groups, and 20 of each sex to the highest concentration group. These extra animals were killed after 52 weeks of daily dietary exposure to DNOC (Broadmeadow, 1991).

No significant difference in the mortality rate was observed among the treated groups, or in comparison to the untreated control group. No clinical signs of adverse effects from the treatment were recorded. In male rats treated at the highest concentration, food consumption was found to be slightly higher than in the controls (+6%) from week 5 onwards. No effect was noted on the body-weight gain of the animals. No significant alterations were recorded in the haematological and biochemical parameters evaluated in the course of the experiment.

Histopathological examination did not reveal any alteration in any organ that could be attributable to an effect of the treatment with DNOC. No increase in the incidence of any type of tumour was recorded (Broadmeadow, 1991). A NOEL of 0.59 mg/kg b.w. per day was determined in males on the basis of increased food consumption, and 5.03 mg/kg b.w. per day in females (highest administered dose).

7.5 Reproduction, embryotoxicity and teratogenicity

7.5.1 Reproduction

Groups of male and female Sprague Dawley CD rats were exposed to technical grade (97.45% pure) DNOC in their diet throughout maturation, mating, gestation and lactation phases of two successive generations (Coles & Brooks, 1997).

The concentrations administered were 15, 30 and 100 mg of DNOC/kg diet. At all dose levels there were no toxicologically meaningful effects on treated adults compared with untreated controls. From the results, the authors concluded that there were no toxicologically significant effects at any dose level for both generations and concluded that the NOEL for adults and reproductive parameters was 100 mg DNOC/kg diet (the highest concentration administered), corresponding to 7.20 mg/kg b.w. per day for F_0 males, 9.24 mg/kg b.w. per day for F_0 females, 10.1 mg/kg b.w. per day for F_1 males and 10.55 mg/kg b.w. per day for F_1 females. However, the reported data indicate that during the gestation phase of the F_0 generations of the 100 mg DNOC/kg diet group there was a statistically significant reduction in group mean body weight ($p < 0.05$) compared to controls from days 7–21 of gestation. No such effect was observed in any other treated group.

During lactation of the F_0 generations the same statistically significant effect, in addition to a statistically significant reduction in food consumption, was noted in the high-dose group only. In the same group (100 mg/kg diet) there was a statistically significant reduction in group mean litter size on days 14 and 21 of lactation ($p < 0.01$) compared to control values in the F_1 generations. Litter size at birth was comparable to controls and the later reduction was mainly due to 3 litters which all had 6 or fewer pups left at days 14 and 21. This effect was also seen in the 30 mg/kg diet group on the same days of the lactation phase ($p < 0.05$).

In the F_0 and F_1 generation 100 mg/kg diet groups, there was a slight, but statistically significant, reduction in group mean litter weight ($p < 0.01$ on days 14 and 21 post-partum). However, the effects seen in the high-dose group of both F_0 and F_1 generations were limited in importance. Therefore, on this basis, the NOEL should be considered to be the intermediate concentration level of 30 mg/kg diet, equivalent to 1.73 mg/kg per day for the F_0 males, 2.24 mg/kg per day for the F_0 females, 2.40 mg/kg per day for the F_1 males and 2.61 mg/kg per day for the F_1 females.

7.5.2 Teratogenicity and embryotoxicity

7.5.2.1 Rat oral study

Groups of pregnant SPF Wistar rats received, through their drinking water, doses equivalent to 1.0, 5.0 or 25.0 mg DNOC/kg b.w. per day on days 6–15 (inclusive) of gestation. No abnormal clinical signs were observed in the pregnant females during the course of gestation. No mortality was recorded. No significant effect was seen on the body-weight gain. At the high-dose level, a reduced food consumption was noted. No stillbirth was recorded. No difference in the rate of resorption was observed, although a slight increase was seen in the high-dose group. Fetal weights were comparable between the different groups and no difference in the incidence of common malformation was recorded. The no-observed-adverse-effect-level (NOAEL) for teratogenicity and for embryotoxicity was determined to be 25 mg DNOC/kg b.w. per day (Dickhaus & Heisler, 1984).

7.5.2.2 Mouse oral study

A daily dose of 8 mg DNOC/kg b.w. was administered orally on days 11, 12, 13 and 14 of gestation; no adverse effect was obtained, whereas the positive control (ethylmethane sulphonate) given as a single dose of 300 mg/kg induced 4% exencephaly and 7% other brain malformations (Nehéz et al., 1981).

7.5.2.3 Rabbit oral studies

Groups of pregnant chinchilla (Kfm: CHIN.SPF) rabbits, 16 females/group, received doses of 0 (controls), 4.0, 10.0, 25 mg DNOC/kg per day by gavage on days 6–18 (inclusive) of gestation.

Mortality was recorded in the high-dose group, in which four pregnant females died. These deaths occurred during the first 5 days and were not treatment-related. Two deaths occurred late in the study (days 26 and 27 of gestation); lung and intestinal infections were present at necropsy.

Laboured respiration in the high-dose group animals was observed. Food consumption was generally increased in the treated groups when compared to that of the controls, with no concomitant meaningful effects observed on the maternal body-weight gain of the treated animals indicating a metabolic stress in relation to the known mechanism of action of DNOC on the pregnant females.

No biologically significant effects were observed in fetal weights. At the high-dose level, 29 fetuses out of a total of 64 in 8 litters out of 10 showed either external or visceral malformations or skeletal variations. The most frequent malformations were microphthalmia or anophthalmia (24 fetuses in 8 litters) and hydrocephaly or microcephaly (21 fetuses in 6 litters). In only one of the affected litters, 6 fetuses out of 8 had multiple malformations (limbs, omphalocele or gastroschisis) in addition to microcephaly or hydrocephaly. The NOAEL was determined at 10 mg/kg b.w. per day for fetal effects, on the basis of the malformations observed in the high-dose group (Allen et al., 1990a).

7.5.2.4 Rabbit dermal studies

Groups of 16 pregnant chinchilla (Kfm: CHIN.SPF) rabbits were treated by daily cutaneous applications of doses of 0 (controls), 10, 30 or 90 mg DNOC/kg b.w. The contact with the skin was maintained for 6 h/day under an occluding bandage, from day 6 to day 18 (inclusive) of gestation.

No irritating effect was observed at the application sites throughout the course of the study. One mortality was recorded in the high-dose group on day 9 of gestation. No effect was seen on the food consumption. At the high-dose level, the body-weight gain decreased during the first week of treatment but was comparable to the controls afterwards.

Two females in the high-dose group had total resorption. No other adverse effects were observed in the maternal parameters. No effect was seen on the fetal body weights. No significant, or dose-related, teratogenic response was observed in this study. Two fetuses in two separate litters in the low-dose group had hydrocephaly or exencephaly. One isolated case of hydrocephaly was seen in the low-dose level and one in the mid-dose group. No teratogenic response was observed in the high-dose group.

The NOELs were determined as 10 mg/kg b.w. per day for embryotoxicity on the basis of the total resorptions seen in two females in the high-dose group, 30 mg/kg b.w. per day for maternal effects on the basis of early effects on the body weights and 90 mg/kg per day for teratogenicity (Allen et al., 1990b).

7.5.2.5 Mouse intraperitoneal studies

Following a non-standard methodology, Nehéz et al. (1981) demonstrated a lack of teratogenic effects of DNOC in DBA mice. The pregnant mice were injected intraperitoneally with a single, or repeated, dose of 15 mg DNOC/b.w. (half of the intraperitoneal LD_{50}) either on day 11, or on days 4, 9 and 11, of pregnancy. The values of post-implantation loss, the mean weight of the embryos and the incidence of malformations did not significantly differ from those obtained in untreated pregnant mice.

7.6 Mutagenicity and related endpoints

The genotoxic potential of DNOC has been investigated in several *in vitro* and *in vivo* test systems (Table 6). Several older experiments, not always performed according to standard practices, have been included.

Table 6. Studies on mutagenicity of DNOC

Test systems	Cells/species/endpoint	Concentrations/doses	Activation	Results	Reference
Microbial systems and lower organisms					
Bacteria	*Proteus mirabilis* repair deficient/repair proficient	10 mg	without	+	Adler et al. (1976)
	Salmonella typhimurium TA1537, TA98, TA100	100 µg/plate	without	−TA1537 −TA98 −TA100	Somani et al. (1981)
	Salmonella typhimurium TA98, TA100	1 µmol/plate	with without	+TA98 −TA100 −TA98 −TA100	Nishimura et al. (1982)
	Salmonella typhimurium TA1535, TA1538, TA98, TA100	·	without	−TA1535 +TA1538 +TA98 +TA100	Sundvall et al. (1984)
	Salmonella typhimurium TA98NR and TA100NR			±TA98NR −TA100NR	

Table 6 (contd)

	Salmonella typhimurium TA1535, TA1537, TA1538, TA98, TA100	5–500 µg/plate	with without with and without	+TA1535 –TA1535 +TA98 +TA1537 +TA1538 –TA100	Marzin (1991a)
	Salmonella typhimurium TA97, TA98, TA100, TA102	3–1000 µg/plate	with and without	–TA97 –TA98 –TA100 –TA 102	Hrelia et al. (1994)
Drosophila melanogaster	Sex-linked recessive lethal	1/3–2 x LC		+	Müller and Haberzetti (1980)
Mammalian cells in vitro					
	HGPRT, mouse lymphoma L5178Y cells	31.3–500 µg/ml	with and without	+ –	Martin (1981)
	HGPRT, Chinese hamster V79 cells	100-300–1000–3000 µg/ml	with without	+ –	Marzin (1991b)
	chromosome damage, Chinese hamster ovary cells	0.25–25 µg/ml	with and without	– –	Garner (1984)
	chromosome damage, human lymphocytes	2 µg/ml 0.02 µg/ml		+ +	Nehéz et al. (1977) Nehéz et al. (1975)

Table 6 (contd)

	chromosome damage, human lymphocytes	100–300–1000–100 µg/ml	with without	–	Marzin (1991d)
	sister chromatid exchange, human lymphocytes	3–10–30 µg/ml 7.66 and 7.51 8.03 and 8.32 µg/ml	with without	– –	Hrelia et al. (1994)
	unscheduled DNA synthesis human lymphocytes	12–25–50–100 µg/ml	without	–	Hrelia et al. (1994)

Mammalian cells in vivo

Somatic cells	bone marrow micronucleus, mouse	20 mg/kg b.w. 10 mg/kg b.w. after 1 year, i.p.		+ +	Nehéz et al. (1984)
	bone marrow chromosomal aberrations, rat	4–16 mg/kg b.w., oral		–	Kirland (1984)
	bone marrow chromosomal aberrations, mice	3–12 mg/kg b.w., i.p.		–	Kirland (1986)
	bone marrow micronucleus, mouse	20 mg/kg b.w., i.p.		–	Marzin (1991c)

Table 6 (contd)

	bone marrow chromosomal aberrations, rat	7.5, 15, 30 mg/kg b.w. i.p.	+	Hrelia et al. (1994)
	DNA unwinding, rat hepatocytes	1–9.3 mg/kg b.w., i.p.	+	Grilli et al. (1991)
	unscheduled DNA synthesis, rat hepatocytes	28, 70 mg/kg b.w. oral	–	Fellows 1998
Germ cells	germ cells dominant lethal assay, mouse	8, 10, 15 mg/kg b.w. i.p.	+	Nehéz et al. (1978)
	germ cells meiotic chromosomal, mouse chromosomal aberration F. embryo, mouse	4 × 5 mg/kg b.w., p.o. to F₀ female	+	Nehéz et al. (1981)
	germ cells meiotic chromosomal, mouse	0.6 mg/kg b.w. per day¹ × 10 days, i.p.	±	Nehéz et al. (1982)
	chromosomal aberrations F₁ embryo, mouse	10 mg/kg b.w. per day to F₀ male; i.p.	+	Nehéz et al. (1984)
	testes wt., sperm count, sperm abnormalities, mouse	3–12 mg/kg b.w. i.p. or oral	–	Quinto et al. (1989)

ᵃ per ml wastewater or per μg test substance ᵇ +, positive results; ±, equivocal results; –, negative results.
HGRPT, hypoxanthine–guanidine phosphoryl transferase ⁽³⁾ Krezonit E containing 50% DNOC

7.6.1 Microbial systems and lower organisms

There was some evidence of genotoxicity in a DNA repair proficient/deficient test in *Proteus miribalis*. There were several studies in *Salmonella typhimurium* with some positive and some negative responses both in the presence and in the absence of metabolic activation. The study by Sundvall et al. (1984) showed positive responses in strains TA98, TA1538 and TA100, and a strong reduction or abolition of the positive response in the nitroreductase-deficient strains TA98NR and TA100NR, suggesting involvement of one of the nitro groups, which would be expected to be reduced by colonic bacteria such as *S. typhimurium*. The negative result of the most recent study by Hrelia et al. (1994) in strains TA97, TA98, TA100 and TA102 is somewhat surprising. There was a positive response in *Drosophila melanogaster* in the sex-linked recessive lethal assay, but doses were very high.

7.6.2 Mammalian cells in vitro

Positive results were found for gene mutation in mouse lymphoma cells at the HGPRT locus (Martin, 1981), but not for Chinese hamster cells (Marzin, 1991b). Negative results were found in human lymphocytes for sister chromatid exchanges and unscheduled DNA synthesis in both the presence and the absence of metabolic activation at doses up to 50 µg/ml (Hrelia et al., 1994). Nehéz et al. (1977, 1978) examined Krezonit E (containing 50% DNOC) for the induction of chromosomal aberrations in cultured human lymphocytes. Very high frequencies were reported at very low concentrations. Perhaps some other component was present in the Krezonit E pesticide that could account for its toxicity and chromosome-breaking ability. There were some increases in chromosome aberrations in other studies (Garner, 1984).

7.6.3 Mammalian cells in vivo

Two studies for chromosomal aberrations in rat and mouse bone marrow have been conducted (Kirkland, 1984, 1986) and one for micronuclei in mouse bone marrow (Marzin, 1991d); all were reported as negative. Increases were seen but were not considered biologically significant. However, there was a positive response in a rat bone marrow aberration study (Hrelia et al., 1994). There was a negative unscheduled DNA synthesis assay (Fellows, 1998) and a positive DNA unwinding study (Grilli et al., 1991). The positive germ cell effects found by Nehéz et al. (1978 and 1981) may have been due to another component of Krezonit E, as discussed above.

In conclusion, some positive responses have been detected in *Salmonella, Drosophila* and mammalian cells *in vitro* and *in vivo*. However, the weight of evidence in studies conducted to Good Laboratory Practice (GLP) standards *in vivo* have generally produced negative responses. On the basis of all the data available, the mutagenicity of DNOC remains equivocal.

7.7 Carcinogenicity

A carcinogenicity study performed in the rat is summarized in section 7.4.1. No increase in the incidence of any type of tumour was recorded (Broadmeadow, 1991).

7.8 Special studies

7.8.1 Cataractogenicity

Diets containing 0.25% DNOC (or 2500 mg/kg feed) were administered for 2 days to 5-day-old white Peking ducklings (Spencer et al., 1948). Cataracts were produced within 24 h in all the birds. This concentration induced 56% mortality the first day and 100% the second day.

7.8.2 Immunotoxicity

Vos et al. (1983) tested DNOC in the category of compounds having no, or only marginal, effects on immunological parameters. Wistar-derived rats received DNOC in their diet at concentrations of 25, 100 and 400 mg/kg of feed for 3 weeks. General toxicological effects were evaluated, in addition to the particular immunological parameters: lymphocyte and monocyte counts; serum IgM and IgG levels; weight and histopathology of thymus, spleen and lymph nodes. None of these parameters was significantly affected. DNOC was not considered as having the potential to induce disturbances in the immunological system.

7.8.3 Testicular toxicity

To assess its potential testicular toxicity, DNOC was administered intraperitoneally or orally to (C3H × C57BL/6) F_1 mice in 10 ml of distilled water at 3, 6 or 12 mg/kg b.w. once a day for 5 consecutive days. Testicular weight and sperm counts were measured in each mouse. DNOC did not affect the testicular weight at any dose by any route, and did not induce any sperm abnormalities (Quinto et al., 1989).

7.9 Factors modifying toxicity; toxicity of metabolites

7.9.1 Factors modifying toxicity

The acute toxicity of DNOC is enhanced by increased environmental temperature. No mortality occurred in 10 rats administered orally a single dose of 50 mg DNOC/kg b.w. and kept for 4 h at a temperature between 20 and 22 °C, whereas 70% of the rats similarly treated, but kept for 5 h at a temperature of 37–40 °C, died (Harvey, 1952). The effect of temperature on the survival of animals dosed with DNOC is very marked. Deaths among groups of rats receiving a series of daily injections of 20 mg DNOC/kg b.w. were reduced from 31% in rats kept in a "warm" laboratory to between 8.5% and 3.5% when the animals were placed in a cool, draughty corridor (Parker et al., 1951). These observations have been confirmed by King & Harvey (1953a), Keplinger et al. (1959) and, in mice, by Tesic et al. (1972).

It was shown that the amount of DNOC absorbed from the gastrointestinal tract of the rat, and the resulting blood level, depend on the type of fat administered shortly after dosing with DNOC (olive oil or castor oil). Gastrointestinal uptake of DNOC increases when 0.2 ml of olive oil is given, whereas 1.0 ml has no influence. Castor oil in a non- purgative dose (0.2 ml) inhibits resorption, and then slows down the decrease of DNOC blood level over a period of 48 h. Rapeseed oil, which is more slowly absorbed than olive oil, shows only a slight inhibitory effect on DNOC digestive absorption (Starek & Lepiarz, 1974).

7.9.2 Toxicity of metabolites

Froslie & Karlog (1970) have demonstrated methaemoglobinaemia in cattle poisoned with DNOC, and have assigned this effect to the further metabolism of the amino compounds into diamino derivatives in the rumen. Diaminophenols are not produced in any significant quantities in non-ruminants (Truhaut & de Lavaur, 1967). It was also demonstrated that 6-ANOC was less toxic than its parent compound, DNOC, by a factor of at least 20 (Smith et al., 1953).

7.10 Mechanisms of toxicity; mode of action

DNOC, like other substituted dinitrophenols, is an uncoupler of mitochondrial oxidative phosphorylation (Judah, 1952; Ilivicky & Casida, 1969; Moreland, 1980). Oxidation and phosphorylation are tightly coupled because oxidation cannot proceed via the respiratory chain without concomitant phosphorylation of adenosine diphosphate (ADP) to adenosine triphosphate (ATP). The rate of mitochondrial respiration is therefore controlled by the concentration of ADP.

By preventing phosphorylation of ADP without interfering with electron transfer, DNOC dissociates oxidation in the respiratory chain from phosphorylation. This results in respiration becoming uncontrolled since the concentrations of ADP and inorganic phosphate no longer limit the rate of respiration. Oxygen consumption and the release of energy in the form of heat therefore increase, because energy is no longer captured by ADP to form ATP. As a consequence the shortage of ATP in critical organs, such as heart and respiratory muscles, may lead to the blocking of their vital functions. Increased oxygen consumption and dissipation of energy in the form of heat represent the hallmark of the pharmacological and toxicological effects of DNOC.

The exaggeration of catabolic processes, involving glycolysis, glycogenolysis and fatty acid metabolism, has been shown to be linearly related *in vivo* in guinea-pigs with the dose of DNOC (Harvey, 1959). The increased oxygen consumption has been shown not only in mammals but also in birds, bees and urchin eggs (Parker et al., 1951). Dissipation of energy may result in elevation of the body temperature, leading eventually to severe hyperthermia.

Shortage of ATP at muscular levels may lead to muscular paralysis and, in the case of death by DNOC poisoning, to early *rigor mortis*.

The pharmacological effect of DNOC, which has been used to induce weight loss associated with normal or increased food intake, results from increased degradation of fatty acids, inhibition of lipogenesis and increased glycolysis and glycogenolysis (Gasiewicz, 1991).

8. EFFECTS ON HUMANS

8.1 General population exposure

8.1.1 Clinical studies

In the 1930s, DNOC was used for a limited period of time as a therapeutic agent to treat obesity (Dodds and Robertson, 1933a,b; Gasiewicz, 1991). The initially recommended daily dose was 3 mg/kg b.w. for 4–5 days. This dosing regimen induced an increased basal metabolic rate (BMR) with sweating, lethargy and restlessness. Lower, but still effective, doses were recommended: 0.5–1.0 mg/kg per day. However, an increase in BMR of 30–50% was observed, together with clinical signs.

8.1.2 Acute toxicity

Signs and symptoms of acute DNOC poisoning in humans include nausea, gastric distress, restlessness, sensation of excessive heat, sweating, thirst, deep and rapid respiration, tachycardia, pyrexia, cyanosis, collapse and coma. Death is promptly followed by intense *rigor mortis*. A hot environment enhances the intensity of the symptoms and shortens the time until their occurrence. Acute DNOC poisoning runs a rapid course, and the general rule is either death or recovery within 24–48 h (Hayes, 1963; Morgan, 1982).

A case of poisoning in a 4-year old child resulting from a dosing error in the preparation of an ointment has been reported. Fifty grams of this ointment, containing 25% DNOC, was applied to the boy's skin. This led to acute poisoning and death in 3.5 h. The applied dose was calculated to be 757 mg/kg b.w. (Gasiewicz, 1991).

8.2. Occupational exposure

It seems that, over the last 25 years, the number of cases of occupational intoxication by DNOC has dramatically declined. This is certainly the result of better education of users, who are now well aware of the toxic properties of the product, and of the importance of wearing adequate protective equipment. The improvement of application techniques, equipment and formulations has also been a determining factor, as was the significant decline in the agricultural use of DNOC.

Several old cases of subacute exposure during manufacturing can be found in the literature. Malter (1949) described the case of six workers exposed to DNOC dust for a month before the first symptoms occurred: general muscular lassitude; nausea; anorexia; profuse sweating; weight loss (an average of 3% of the initial body weight, with one subject losing 2 kg in 48 h) and insomnia. There was neither hyperthermia nor polypnoea. The workers were removed from exposure without any treatment, except for a high-calorie diet for 2 weeks. All signs subsided, apart from persistent neuritic-type pain in the legs in one subject. No indications of exposure doses, or levels of DNOC in the blood, were provided in the report.

Malter (1949) also mentioned several cases from the literature: four fatal cases in Germany, two in the Netherlands reported in 1946 and two cases in England in 1947. Bidstrup & Payne (1951) reported eight fatal cases of subacute poisoning that occurred in the United Kingdom (one manufacturing, and seven agricultural workers) between 1946 and 1950. The authors observed that: "both in Great Britain and in other countries, all the cases of fatal poisoning have occurred during unusually hot weather". All the clinical symptoms and necropsy findings described in the paper are consistent with those of DNOC poisoning. These are briefly described in the WHO (1982) review. An extensive analysis of the signs and manifestations of DNOC poisoning, as well as of their circumstances and conditions of occurrence, has been made by Hunter (1953) and by Stott (1953).

A recent field exposure study has been performed in the Netherlands on 11 male agricultural workers: 4 were private farmers, 6 contract sprayers and 1 truck driver carrying the chemical from the warehouse to the fields (Heuts, 1993). DNOC was sprayed as an oily formulation, containing 200 g a.i./litre, used in July and August for desiccating the haulm of seed potatoes before harvest. Blood samples were taken at regular intervals from the workers and DNOC levels were determined, together with several liver function parameters. The total exposure time during the spraying season varied from 4 to 20 h per person for the individual farmers, and from 2 to 22 h for the professional sprayers. The total quantity of formulated product sprayed during the 2-week period before the blood sampling ranged from 100 to 400 litres for the farmers, and from 145 to 2193 litres for the contractors.

DNOC concentrations in the blood ranged from <0.5 mg DNOC/litre of blood to 0.6 and 0.8 mg/litre in two farmers, respectively. Less than 0.5 mg DNOC/litre of blood was detected in professional sprayers (the limit of detection of the analytical method used was 0.5 mg/litre). No biological alterations of the liver parameters were detected and no complaints of clinical signs of exposure were recorded during the study. This study shows that, under present agricultural use conditions, the exposure of sprayers is kept to low levels and is unlikely to induce acute toxic effects.

In 1956 Stott described two cases of polyneuritis observed among staff servicing aircraft that were used for spraying DNOC on locust swarms in eastern Africa (Stott, 1956). The product used was a 20% solution of DNOC in oil. Both men were heavily exposed and did not wear protective equipment. Skin was considered to be the main route of contamination, and both patients displayed marked yellow staining from DNOC both on the palms of their hands and on the sides of their feet. Serum DNOC levels in these patients were 7.6 1g/ml 1 week after the end of exposure and 11.5 1g/ml at the end of exposure, respectively.

The duration of exposures before they were seen by a physician were 2 weeks and 17 days, respectively. Both patients complained of sensations of pins and needles and tingling on the back of the hands and fingers, and numbness in the legs. One had evidence of excessive sweating in the arms and legs; the other also had a loss of sensation to pinprick and cotton wool on the back of his fingers and the dorsal aspects of his toes. There was no loss of motor function in either patient. Signs and symptoms disappeared 12 and 7 days, respectively, after the end of exposure to DNOC.

The characteristics of the clinical picture, the lack of signs of systemic poisoning, the rapid recovery after removal from exposure, the heavy contamination of hands and feet and the lack of correlation with serum DNOC levels suggest a local effect of DNOC.

A fatal case was reported by Steer (1951) where a 21-year-old man was brought to the hospital having felt sick after spraying DNOC for several days. His body had heavy yellow staining. The clinical description of the symptoms was consistent with DNOC poisoning. He had a sudden generalized convulsion with tonic spasms. After his heart fibrillated, the patient died. Shortly after death, postmortem lividity was noticed around the neck and *rigor mortis* was well marked within 45 min. The main finding at necropsy the next day was that the blood still had not clotted. A blood sample taken immediately after death contained 75 µg DNOC/g, whereas a sample taken at necropsy showed only 4.3 µg/g and DNOC levels in other tissues were comparable with the latter value.

A limited number of cases of poisoning, with hepatic alterations, have been reported following either agricultural or industrial exposure. In the case reported by Prost et al. (1973), a farmer, without any unusual history of health problems, felt sick after spraying his fruit trees with DNOC during the afternoon, without protective equipment. His face, hands, hair and work clothes were stained yellow and he complained of a headache. The next morning he was sweating profusely and had a fever of 39 °C. He sought medical attention, and the biological evaluation revealed a significant liver effect with increased bilirubin, increased activity of alanine aminotransferase (ALT; formerly known as SGOT) and aspartate aminotransferase (AST; formerly known as SGPT), and decreased cholesterol. During the following 2 months, the values progressively returned to normal.

A few comparable cases have been reported from Belgium (Herman & Heyndrickx, 1957; Heyndrickx et al., 1962; Heyndrickx et al., 1964). Necropsy findings and DNOC levels in the tissues were reported. Gaultier et al. (1974) have reported three cases of manufacturing workers exposed to DNOC for periods of 2–8 weeks before becoming ill. One case was fatal. The case with the longest exposure showed limited and transient kidney and liver effects, in addition to the usual symptomatology.

Similarly, Jastroch et al. (1978) described the case of a young farmer exposed for a total of 70 h. The description of the case indicates that the patient had a slight kidney and liver impairment. However, it is not known whether this condition existed before the poisoning, or was a consequence of it. At the time of admission to hospital, the blood level was 70 µg DNOC/ml, and the clinical symptomatology was typically that of subacute DNOC poisoning. A liver biopsy performed 14 days later showed some parenchymal disturbances which were not described.

In a review of pesticide poisoning cases in agricultural workers during the year 1991 by the National Dutch Poison Control Centre, van der Laar et al. (1993) identified one case of a local irritation and two others of systemic intoxication by DNOC. All three cases resulted from either technical imperfections or careless handling of the products (inadequate protective clothing).

In the two systemic cases, one was mild, causing only complaints of a feeling of heat, headache and malaise. The second case was more severe and occurred in a worker exposed for 3 weeks. This professional applicator complained of fever, deep sensation of malaise, shortness of breath and body-weight loss. However, the DNOC plasma level found was 0.8 µg/ml, which seems much too low to be responsible for such a typical picture of intoxication, since it appears from many data quoted in the literature that symptoms become apparent only when the concentration in the blood reaches 40 µg/ml.

Table 7 summarizes the measured blood levels and effects in humans after exposure to DNOC. No effects are seen when the blood level of DNOC is below 20 µg/g blood. (For details, see section 8.2.).

In their recommendation of health limits for pesticides, WHO (1982) stated that on the basis of human data a blood level below 20 µg/g will probably not lead to manifest health impairment.

The Dutch Expert Committee on Occupational Standards (DECOS), in their report prepared on behalf of the Industrial Medicine and Hygiene Unit of the Health and Safety Directorate of the Commission of the European Communities, recommended, based on human exposure data, a biological operation exposure limit (BOEL) of 10 mg/g in serum and 5 µg/g in whole blood for workers not exposed to heat stress (Jongerius & Jongeneelen, 1991).

The Task Group therefore concludes, on the basis of human data, that a blood DNOC level below 20 µg/g of whole blood will probably not lead to manifest health impairment.

Table 7. Measured blood levels in humans and effects after exposure to DNOC

Blood level (µg/g)	Comments	Reference
75	Fatal case. Blood sample taken immediately after death. Yellow staining all over the body. Blood level at necropsy 4.3 µg/g	Steer (1951)
70	Farmer exposed totally 70 h. Clinical symptomatology typical of subacute DNOC poisoning. Slight kidney and liver impairment	Jastroch et al. (1978)
48	One volunteer 75 mg/day, 7 days. No symptoms of poisoning	Harvey et al. (1951)
20	Volunteers 75 mg/day, 5 days. No symptoms of poisoning	Harvey et al. (1951)
0.8	One professional applicator exposed 3 weeks. Deep sensation of malaise, shortness of breath, body-weight loss	van der Laar et al. (1993)
<0.5–0.8	Farmers spraying 4–20 h. No biological alterations of liver parameters. No complaints of clinical signs.	Heuts (1993)
<0.5	Professional sprayers 2–22 h. No biological alterations of liver parameters. No complaints of clinical signs. 0.5 was limit of detection in method used	Heuts (1993)

9. EFFECTS ON ORGANISMS IN THE LABORATORY AND FIELD

9.1 Micro-organisms

Growth of three strains of *Rhizobium leguminosarum* was stimulated by a concentration of 250 mg DNOC/litre of culture medium (Tewfik & Evans, 1966).

9.2 Aquatic organisms

The acute toxicity of DNOC to aquatic organisms is summarized in Table 8. LC_{50} values range from 0.066 to 5.7 mg/litre with considerable variation even within animal groups.

Early life stage exposure of embryos of the common carp (*Cyprinus carpio*) was carried out at 4 concentrations of DNOC (0.25, 0.5, 1.0 and 2.0 mg/litre) at 3 pH values (6.9, 7.8 and 9.0) over a 13 day period from fertilization. Mortality decreased markedly with increasing pH. The no observed effect concentrations (NOECs) of DNOC are at or below 0.25 mg/litre at pH 6.9, and 0.5–1 mg/litre at pH 7.8. At pH 9.0, no toxic effect was observed at the highest concentration (2.0 mg/l) (Ghillebaert et al., 1995).

9.3 Terrestrial organisms

9.3.1 Earthworms

The acute effects of DNOC on the earthworm species *Eisenia fetida* were determined according to OECD protocol 207. The LC_{50} (7 days) and LC_{50} (14 days) were found to be 17 mg and 15 mg DNOC/kg of soil, respectively. The NOEC (14 days) was 10 mg DNOC/kg soil (van der Hoeven, 1992).

9.3.2 Honey bees

The LD_{50} (oral) and LD_{50} (contact) for honey bees (*Apis mellifera*) were reported to be 2.04 ± 0.25 µg DNOC/bee and 406 ± 27 µg DNOC/bee, respectively (Beran & Neururer, 1955).

Table 8 Acute toxicity of DNOC to aquatic organisms

Organism	Endpoint	Concentration (mg/litre)	Reference
Micro-organisms			
Bacterium			
Pseudomonas putida	toxic threshold 16 h EC_s (growth)	16	Bringmann & Kühn (1980)
Cyanobacterium			
Microcystis aeruginosa	toxic threshold 72 h EC_s (growth)	0.15	Bringmann & Kühn (1978)
Green algae			
Scenedesmus quadricauda	toxic threshold 7 d EC_s (growth)	13	Bringmann & Kühn (1980)
Scenedesmus subspicatus	96 h EC_{50} (biomass)	6	Sewell et al. (1995a)
	48 h EC_{50} (growth rate)	12	Sewell et al. (1995a)
Protozoans			
Entosiphon sulcatum	toxic threshold 72 h EC_s (growth)	5.4	Bringmann & Kühn (1980)
Chilomonas paramecium	toxic threshold 72h EC_s (growth)	5.4	Bringmann & Kühn (1981)
Uronaemia parduczi	toxic threshold 72 h EC_s (growth)	0.012	Bringmann & Kühn (1981)
Invertebrates			
Water flea			
Daphnia magna	24 h LC_{50}	5.7	van der Hoven (1984)
	14 d LC_{50}	1.6	van der Hoven (1984)
	14 d NOEC (reproduction)	0.6	van der Hoven (1984)
	24 h LC_{50}	2.3	Kühn et al. (1989)
	24 h NOEC (mortality)	1.5	Kühn et al. (1989)
	21 d NOEC (reproduction)	1.3	Kühn et al. (1989)
Daphnia pulex	48 h LC_{50}	0.145	Mayer & Ellersieck (1986)

Scud			
Gammarus fasciatus	96 h LC₅₀	0.11	Mayer & Ellersieck (1986)
Insect			
Pteronarcys californica	96 h LC₅₀	0.32	Mayer & Ellersieck (1986)
Vertebrates (fish)			
Bluegill sunfish			
Lepomis macrochirus	96 h LC₅₀	0.95	Sewell et al. (1995b)
	96 h LC₅₀	0.36	Mayer & Ellersieck (1986)
Rainbow trout			
Oncorhynchus mykiss	96 h LC₅₀	0.45	Sewell et al. (1995c)
	96 h NOEC	0.32	Sewell et al. (1995c)
	96 h LC₅₀	0.066	Mayer & Ellersieck (1986)

Jones & Edwards (1952) reported the results of their experimental work on honey bees, both in the laboratory and in the field, under a range of varied conditions of temperature, humidity and formulations. Although DNOC was toxic to honey bees under laboratory conditions, in the field the repellent effect of the sprayed solutions, followed by the wilting and shrivelling effect, made sprayed flowers unattractive to bees. Below a certain temperature, the bees do not forage and DNOC is used in the autumn and winter; this combination of factors makes it unlikely that bees will forage on treated plants, thus reducing the risk of significant exposure and toxicity.

9.3.3 Birds

The oral acute toxicity of DNOC was determined in the Japanese quail (*Coturnix japonica*). DNOC was administered by gavage as a single dose of 10–25 mg/kg b.w. The LD_{50} (24 h) was calculated as 14.8 mg DNOC/kg b.w. (95% confidence interval 13–17) (Dickhaus & Heisler, 1980). Acute oral LD_{50} values for pheasant and mallard duck were 31.8 and 22.7 mg/kg b.w., respectively (Hudson, et al., 1984).

The subacute dietary toxicity was evaluated by feeding Japanese quail a diet containing concentrations of 555–960 mg DNOC/kg of feed for 5 consecutive days, followed by 3 days' observation period. The LC_{50} (8 days) was determined to be 637 mg DNOC/kg of diet, equivalent to approximately 106 mg DNOC/kg b.w. (Til & Kengen, 1980).

Additional data on the acute toxicity of DNOC to wild birds have been reported by Janda (1970). Pheasants and partridges were treated with one gavage of a water dilution of a formulated product containing 35% DNOC. The LD_{50} values were 24 mg of formulation/kg b.w. (equivalent to 8.4 mg DNOC/kg b.w.) for pheasants, and 23.7 mg of formulation/kg b.w. (equivalent to 8.3 mg DNOC/kg b.w.) for partridges. In a subchronic study reported in the same paper, in which the product was administered by gavage during 3 consecutive days, the LD_{50} values were 7.1 DNOC/kg b.w. per day and 11.1 mg DNOC/kg b.w. per day for pheasants and partridges, respectively. These results indicate that DNOC does not accumulate in these birds. Some repulsive effect was observed when maize, barley and wheat, treated with a concentration equivalent to that sprayed in normal field use, were offered to the birds in a "no-choice" test (i.e., birds were offered only treated grain).

9.3.4 Other wildlife species

Effects on wild animals have been evaluated on the basis of experimental data and field surveys. In agricultural use, DNOC has only very rarely been found to be responsible for poisonings. For example, only three dead hares in which DNOC was found in the carcasses were reported in France over a 3-year period (Grolleau, 1967; Quidet, 1975; de Lavaur & Arnold, 1977, 1981). Grolleau (1967) has determined experimentally that, in the field, a residue level of 500 mg/kg on maize represented a threshold below which no accidental poisoning was expected in hares. Similarly, de Lavaur (1967) demonstrated under actual field conditions that no mortality was observed in treated fields in which up to 964 mg/kg of DNOC were detected in the treated wheat.

10. EVALUATION OF HUMAN HEALTH RISKS AND EFFECTS ON THE ENVIRONMENT

10.1 Evaluation of human health risks

The main exposure for humans is through occupational activities during manufacturing and agricultural use. No exposure of the general population is likely to occur, because the particular use patterns of the end products do not result in residues being present in food. DNOC is an uncoupler of oxidative phosphorylation, and toxic effects observed either in humans or in animals result from this mechanism.

A relatively small number of human cases of acute poisoning have been reported in the literature during both manufacturing and agricultural occupational exposures. In all occasions these occurred from poor hygienic conditions.

The main route of exposure was through skin contact. These cases were reported more frequently several decades ago (see section 8.2). It was noticed that exposure during a period of hot weather increased the risk of acute poisoning. In today's agricultural practices, exposure has been severely reduced through better education of farmers, making them aware of the toxicity of the product and of the necessity to wear personal protective clothing, and to spray with tractors equipped with closed cabins; spraying equipment, formulation and packaging have also been improved.

The onset of clinical symptoms occurred after a relatively short period of exposure. Symptoms included nausea, gastric distress, restlessness, sensation of heat, excessive sweating, thirst, deep respiration, tachycardia and hyperpyrexia. In severe cases, collapse, coma and death occurred within 24–48 h.

A small fraction of the general population was exposed for a limited period of time more than 50 years ago when DNOC was used as a therapeutic agent against obesity. Doses of 3 mg/kg b.w. per day induced an excessive metabolic rate with associated symptoms, including sweating, lethargy and sleep disturbances.

A recent, limited scale, field worker exposure study has shown that the plasma concentration did not exceed 0.8 mg/litre (in an isolated case) and even remained below 0.5 mg/litre (limit of detection) in the majority of workers.

The Group endorses the previous WHO conclusion for a biological threshold limit of 20 µg/ml of whole blood for DNOC (WHO, 1982). A TWA (mg/m^3 per 8 h) of 0.2 has been established in several countries (see section 5.3).

In animal studies, the oral LD_{50} values, as determined in several species, range from 16 to 100 mg/kg b.w. Some positive responses have been detected in *Salmonella*, *Drosophila* and mammalian cells *in vitro* and *in vivo*. The weight of evidence in studies conducted to GLP standards *in vivo* has generally produced negative responses. On the basis of all the data available, the mutagenicity of DNOC remains equivocal. A rat carcinogenicity study did not show any carcinogenic effect. Teratogenicity studies performed in rats and rabbits by oral and percutaneous administration showed that DNOC induced embryotoxicity and teratogenic effects only at dose levels that were also maternally toxic.

10.2 Evaluation of effects on the environment

DNOC has a low vapour pressure and is moderately soluble in water; its application as a pesticide therefore results in distribution to land and surface water with minimal volatilization to the atmosphere. Low concentrations in rainwater may derive from other sources.

Adsorption of undissociated DNOC to particulates in soil is high at low pH (4–5.5); adsorption of dissociated anions is much lower at higher environmentally significant pH (5.5–8) with commonly more than half of the DNOC in the aqueous phase. Adsorption to soil is concentration-dependent, with reduced sorption within the range expected from recommended application rates. Leaching appears to be limited in practice, with occasional low concentrations detected in groundwater.

A range of micro-organisms capable of degrading DNOC has been identified in soils. Half-life in surface waters is of the order of 3–5 weeks and DT_{50} values for soil have been measured at 4–15 days; DT_{100} for soil ranges from 2.5 to 62 days.

EHC 220: Dinitro-ortho-cresol

Use of the compound as a desiccant on potato and an insecticide on dormant fruit crops limits exposure of many organisms. DNOC used for locust control is likely to be sprayed over wide areas; however, details of this usage were not available and no risk assessment can be made.

DNOC is acutely toxic to honey bees. Exposure from winter use would not occur. Desiccant use would be on potato haulms past the flowering stage, and desiccant-treated foliage would not be attractive to bees. Even assuming overspray, the bee hazard quotient ([application rate in g a.i./ha]/[toxicity in µg/bee]) gives no cause for concern.

Little or no exposure of birds and mammals is expected from direct use of DNOC. Indirect exposure from contaminated earthworms is a possibility, but no risk assessment can be made because no residue information is available. Rapid biodegradation could be expected to reduce indirect exposure. Data from a wildlife incident scheme suggest very limited kills of wild mammals. The means of exposure of these animals is unknown.

Earthworms would be exposed to spray reaching the ground and desiccated haulms on the soil surface. Winter use is unlikely to expose worms. Calculated soil concentrations of 7.5 mg/kg are obtained following spraying at 5.6 kg a.i./ha assuming even distribution in the top 5 cm of soil (EPPO/COE Guidelines, EPPO, 1993). This gives a toxicity exposure ratio (TER) of 2.01, regarded as being of moderate concern.

Calculated water concentration following application as a desiccant at 5.6 kg a.i./ha is 0.093 mg/litre (EPPO/COE Guidelines, EPPO, 1993) assuming a level of spray drift of 5% at 1 m from the spray boom (Ganzelmeier et al., 1995). TER values for aquatic organisms are 0.7 for fish (based on rainbow trout), 1.6 for invertebrates (based on *Daphnia magna*) and 64.3 for algae. A spray buffer zone of 5 m from water courses increases the TER values to acceptable levels (59, 130, 5360 respectively). Application to dormant fruit crops is expected to be by drench with large droplet size; aquatic concentrations are assumed to be no higher than for boom application.

It can be concluded that DNOC could adversely affect organisms in the environment following acute exposure; chronic effects are not expected given the degradation of the compound. Field studies are not available for the organisms most likely to be exposed.

11. PREVIOUS EVALUATION BY INTERNATIONAL BODIES

DNOC was evaluated by the FAO/WHO Joint Meeting on Pesticide Residues (JMPR) in 1963 and 1965. No ADI was established.

DNOC is classified in class Ib, 'highly hazardous', in the WHO Recommended Classification of Pesticides by Hazards (WHO, 1999).

REFERENCES

Adler B, Braun R, Schoneich J, & Bohme H (1976). Repair-defective mutants of *Proteus mirabilis* as a pre-screening system for the detection of potential carcinogens. Biol Zbl, **95**:463–469.

Alber M, Boehm HB, Brodesser J (1989) Determination of nitrophenols in rain and snow. Fresenius J Anal Chem, **334**:540–545.

Allen PA, Biedermann K, & Terrier C (1990a) Embryotoxicity study (including teratogenicity) with DNOC Technical in the rabbit (dermal application). (RCC study no. 215638) Itingen, Switzerland (unpublished report prepared for Pennwalt Holland bv).

Allen PA, Biedermann K, & Terrier C (1990b) Embryotoxicity study (including teratogenicity) with DNOC Technical in the rabbit (oral administration). (RCC study no. 215651) Itingen, Switzerland (unpublished report prepared for Pennwalt Holland bv).

Ambrose AM (1942) Some toxicological and pharmacological studies on 3,5-dinitro-*o*-cresol. J Pharm Exp Ther, **76**: 245–251.

Arustamyn AN (1972) [The toxicity of dinitro-*ortho*-cresols for warm-blooded animals and problems of industrial hygiene in its application.] Tr Inst Vet Sanit, **45**:166–169 (in Russian).

Batchelor GS, Walker KC, & Elliott JW (1956) Dinitroorthocresol exposure from apple-thinning sprays. Arch Indust Health, **13**:593–596.

Benadikova H, & Kalvoda R (1984) Adsorptive stripping voltammetry of some triazine- and nitro-group-containing pesticides. Anal Lett, **17**:1519–1531.

Ben Dyke R, Sanderson DM, & Noakes DN (1970) Acute toxicity data for pesticides. World Rev Pest Contr, **9**(3):119–127.

Beran F, & Neururer J (1955) [About understanding of the effects of pesticides on the honey bee (*Apis mellifera* L.) 1. Hazards of pesticides to honey bees.] Pflanzenschutzberichte, **15**(8/12):97–147 (in German).

Bidstrup PL, & Payne DJH (1951) Poisoning by dinitro-*ortho*-cresol. Report of eight fatal cases occurring in Great Britain. BMJ, **ii**:16–19.

Bieber WD (1995) Degradation and metabolism of DNOC in soil (SGS Natec Institut, Hamburg Germany study no. 92 9633, unpublished report prepared for Elf Atochem Agri SA).

Bringmann G, & Kühn R (1978) Limiting values for the noxious effects of water pollutants material to blue-green algae and green algae. vom Wasser, **80**:45–60.

Bringmann G, & Kühn R (1980) Comparison of the toxicity thresholds of water pollutants to bacteria, algae and protozoa in the cell multiplication inhibition test. Water Res, **14**:231–241.

Bringmann G, & Kühn R (1981) [Comparison of the effects of harmful substances on flagellates and ciliates, as well as on holozoic bacteriophagic and saprozoic protozoa.] gwf-wasser/abwasser, **122**:308–313 (in German).

References

Broadmeadow A (1988) Technical DNOC: Preliminary toxicity study by dietary administration to F-344 rats for six weeks (Life Science Research study no. 87/PTN 001/433). Eye, UK (unpublished report prepared for Pennwalt Corporation Agrichemicals Division).

Broadmeadow A (1991) Technical DNOC: Combined oncogenicity and toxicity study by dietary administration to F-344 rats for 104 weeks (Life Science Research study no. PTN/003/DNOC). Eye, UK (unpublished report prepared for Pennwalt Corporation).

Buchholz KD, & Pawliszyn J (1993) Determination of phenols by solid-phase microextraction and gas chromatographic analysis. Environ Sci Technol, **27**: 2844–2848.

Burkatskaya EN (1965a) [Maximum permissible concentration of dinitro-*o*-cresol in air.] Gig Sanit, **30**:34–37 (in Russian).

Burkatskaya EN (1965b) The toxicity of dinitro-*ortho*-cresols for warm-blooded animals and problems of industrial hygiene in its application. Gig Tr Prof Zobal, **9**(4):56–57.

Coles RJ, & Brooks PN (1997) Technical DNOC: dietary two generations reproduction study in the rat (Safepharm Laboratory project no.764/010). Derby, UK (unpublished report prepared for Elf Atochem Agri SA).

de Lavaur E (1967) Résidus de dinitro-4,6-orthocrésol sur blé après différents types de traitement. Phytiat Phytopharm, **16**:15–21.

de Lavaur E, & Arnold A (1977) Pesticides et faune sauvage: Résultats des analyses toxicologiques effectuées sur le gibier de 1974 à 1976. Phytiat Phytopharm, **26**:159–168.

de Lavaur E, & Arnold A (1981) Pollution des vertébrés terrestres sauvages par les pesticides et les PCB. Enquête 1977–1979. Phytiat Phytopharm, **30**:89–95.

den Tonkelaar EM, van Leeuwen FXR, & Kuiper C (1983) Semichronic toxicity of DNOC in the rat. Med Fac Landbouww Rijksuniv, **48**(4):1015–1022.

Dey-Hazra A, & Heisler E (1981) [Acute toxicity of trifocide 50% flowable, 50% DNOC ammonium salt in water by inhalation in the rat.] Pharmatox study no. 1-4-246–81 (unpublished report prepared for Elf Atochem Agri bv) (in German).

Dickhaus S, & Heisler E (1980) [Acute toxicity of technical active substance DNOC (99% ± 1%) after oral administration to quail.] Pharmatox study no. 1-8-239–80. Pharmatox, Hannover, Germany (unpublished report prepared for Ruhr-Stickstoff AG) (in German).

Dickhaus S, & Heisler E (1984) [Teratogenic/embryotoxic study with the product "Trifocide liquid 50%" following oral administration in the rat.] (Pharmatox study no. 2-4-240–83) Hamburg, Germany (unpublished report prepared for Pennwalt Holland bv) in German.

Di Corcia A, & Marchetti M (1992) Method development for monitoring pesticides in environmental waters. Liquid–solid extraction followed by liquid chromatography. Environ Sci Technol, **26**: 66–74.

Diepenhorst PC, Kool P, & Luijdendijk HM (1995) DNOC. Determination by HPLC in human blood. Rotterdam Elf Atochem Agri bv Analytical Dept (unpublished report no. Anal. Meth. SOP DLA-011.7).

Dodds EC, & Robertson JD (1933a) The clinical applications of dinitro-*o*-cresol. Lancet, ii:1137–1139.

Dodds EC, & Robertson JD (1933b) The clinical applications of dinitro-*o*-cresol. II. A study of myxoedema. Lancet, ii:1197–1198.

Driscoll R (1995a) Technical DNOC: Acute oral toxicity test in the rat (Safepharm Laboratories study no. 765/4). Derby, UK (unpublished report prepared for Elf Atochem Agri SA).

Driscoll R (1995b) Technical DNOC: Acute dermal toxicity (limit test) in the rat (Safepharm Laboratories study no. 764/5). Derby, UK (unpublished report prepared for Elf Atochem Agri SA).

Driscoll R (1995c) Technical DNOC: Acute dermal irritation test in the rabbit (Safepharm Laboratories study no. 764/30). Derby, UK (unpublished report prepared for Elf Atochem Agri SA).

Driscoll R (1995d) Technical DNOC: Acute eye irritation test in the rabbit (Safepharm Laboratories study no. 764/31). Derby, UK (unpublished report prepared for Elf Atochem Agri SA).

Driscoll R (1995e) Technical DNOC: Magnusson and Kligmann maximisation study in the guinea pig (Safepharm Laboratories study no. 764/7). Derby, UK (unpublished report prepared for Elf Atochem Agri SA).

EC (1999) *O J EC*, 2 March 1999, Commission decision 1999/164/EC.

EPPO (1993) European Plant Protection Organization. Decision-making scheme for the environmental risk assessment of plant protection products. Chapters 6–11. EPPO Bulletin, 23.

Fabreguettes C (1993) Pharmacokinetic study after single cutaneous application in rats (Centre International de Toxicologie, Evreux, France, study no. 10530/PAR) (unpublished report prepared for Elf Atochem Agri SA).

FAO/WHO (1964) Evaluation of the toxicity of pesticides residues in food; report of a Joint Meeting of the FAO Committee on Pesticides in Agriculture and the WHO Expert Committee on Pesticide Residues. Geneva, World Health Organization. FAO Meeting Report no. PL/1963/13; WHO/Food Add. 23.

FAO/WHO (1965) Evaluation of the toxicity of pesticide residues in food. FAO Meeting Report no. PL/1965/10/1; WHO/Food Add./27–65.

Farrington DS, Martindale RW, & Woolam CJ (1982) Determination of the active ingredient content of technical and formulated dinobuton, dinoseb, dinoterb and DNOC by high-performance liquid chromatography. Analyst, **107**:71–75.

Fellows M (1998) DNOC technical measurement of unscheduled DNA synthesis in rat liver using an in vivo/in vitro procedure. Report No. 160711-DS140. unpublished report by Covance Laboratories Limited, North Yorkshire, UK Prepared for Elf Atochem Agri SA.

References

Froslie A (1973) Methaemoglobin formation by diamino metabolites of DNOC and DNBP. Acta Pharmacol Toxicol, **32**:257–265.

Froslie A, & Karlog O (1970) Ruminal metabolism of DNOC and DNBP. Acta Vet Scand, **11**:114–131.

Ganzelmeier H, Spangenberg R, Streloke M, Hermann M, Wenzelburger H-J, & Walter H-F (1995). Studies on the spray drift of plant protection products. Berlin, Blackwell Wissenschafts-Verlag.

Garner RC (1984) Study to evaluate the chromosome damaging potential of DNOC by its effects on cultured Chinese hamster ovary cells using an in vitro cytogenetics assay. (Microtest Research study no. PHARM 1/CYT/RCG 2) UK (unpublished report prepared for Pennwalt Holland bv).

Gasiewicz TA (1991) Nitro compounds and related phenolic pesticides. In: Hayes WJ Jr, & Laws ER Jr ed. Handbook of pesticide toxicology. San Diego, CA, Academic Press, vol. 3, pp. 1191–1269.

Gaultier M, Gervais P, & Conso F (1974) Intoxication aiguë par le dinitroorthocresol. J Eur Tox, **7**(1):9–11.

Ghillebaert F, Chaillou C, Deschamps F, & Roubaud P (1995) Toxic effects at three pH levels of two reference molecules on common carp embryo. Ecotoxicol Env Safety, **32**:19–28.

Grilli S, Ancora G, Valenti AM, Mazzullo M, & Colacci A (1991) In vivo unwinding fluorometric assay as evidence of the damage induced by fenarimol and DNOC in rat liver DNA. J Tox Environ Health, **34**:485–494.

Grolleau G (1967) Intoxication experimentale de lièvres par le DNOC. Phytiat Phytopharm, **16**:23–25.

GTZ, German Agency for Technical Cooperation (1997) The work of other agencies and organizations relating to prevention and disposal of obsolete and unwanted pesticide stocks. In: Prevention and disposal of obsolete and unwanted pesticide stocks in Africa and the Near East. Second Consultation Meeting. FAO Pesticide Series, **5**:17–21.

Gustafson DI (1989) Ground water ubiquity score: a simple method for assessing pesticide leachability. Envir Tox Chem, **8**:339–357.

Hamdi YA, & Tewfik MS (1970) Degradation of 3,5-dinitro-*o*-cresol by *Rhizobium* and *Azobacter* spp. Soil Biol Biochem, **2**:163–166.

Hallberg GR (1989) Pesticide pollution of groundwater in the humid United States. Agricol Ecosyst Environ, **26**:299–367.

Harvey DG (1952) The toxicity of the dinitro-cresols. J Pharm Pharmacol, **4**:1062–1066.

Harvey DG (1958) Some aspects of the metabolism of 4:6-dinitro-*o*-cresol (DNC) by the ruminant. J Comp Pathol, **64**:54–63.

Harvey DG (1959) The quantitative response of the oxygen consumption and weight of guinea pig to some metabolic stimulants. J Pharm Pharmacol, **11**:681–688.

Harvey DG, Bidstrup PL, & Bonnel JAL (1951) Poisoning by dinitro-*ortho*-cresol. Some observations on the effects of dinitro-*ortho*-cresol administered by mouth to human volunteers. BMJ, 2:13–16.

Hawley GG (1981) The condensed chemical dictionary. 10th ed. New York, Van Nostrand Reinhold.

Hayes WJ Jr (1963) Clinical handbook on economic poisons. Washington, DC, US Department of Health, Education and Welfare (Public Health Service publication no. 476).

Heimlich F, & Nolte J (1993) Determination of the pK values of 2,4-dinitrophenol herbicides using UV spectroscopy. Science Total Envir, **132**:125–131.

Herman M, & Heyndrickx A (1957) [Toxicological investigation of poisoning cases with dinitro-*ortho*-cresol (DNOC) with fatal outcome.]. Med Landb Hogeschool Gent, **22**:647–653 (in Dutch).

Herterich R (1991) Gas chromatographic determination of nitrophenols in atmospheric liquid water and airborne particulates. J Chromatogr, **549**: 313–324.

Heuts FJM (1993) Biological monitoring of DNOC in applicators after use as haulmkiller in seed potatoes. Rotterdam (unpublished report prepared for Elf Atochem Agri bv).

Heyndrickx A, Maes R, & Tyberghein F (1964) Fatal intoxication by man due to dinitro-*ortho*-cresol (DNOC) and dinitrobutylphenol (DNBP). Med Landb Hogeschool Gent, **29**: 1189–1197.

Heyndrickx A, Avermaete M, Maes R, & Schauvliege F (1962) Determination of dinitro-*ortho*-cresol (DNOC) in post mortem material of a farmer. Med Landb Hogeschool Gent, **27**:955–963.

Hope RL, Mullee DM, & Bartlett AJ (1995) Analytical DNOC and technical DNOC. Determination of general physico-chemical properties (Safepharm Laboratories study no. 764/001) Derby, UK, Safepharm Analytical Chemistry Laboratory (unpublished report prepared for Elf Atochem Agri SA).

Hopper ML, McMahon B, & Griffith KR (1992) Analysis of fatty and nonfat foods for chlorophenoxy alkyl acids and pentachlorophenol. J Am Org Anal Chem Int, **75**: 707–713.

Howarth R, Tremain SP, & Bartlett AJ (1995) Technical DNOC (batch number 547-108). Determination of vapour pressure. Safepharm Laboratories study no. 764/002) Derby, UK, Safepharm Analytical Chemistry Laboratory (unpublished report prepared for Elf Atochem Agri SA).

Hrelia P, Vigagni F, Maffei F, Morotti M, Colacci A, Perocco P, Griili S, & Cantelli-Forti G (1994) Genetic safety evaluation of pesticides in different short-term tests. Mut Res, **321**:219–228.

Hudson RH, Tucker RK, & Haegele MA (1984) Handbook of toxicity of pesticides to wildlife. US Department of the Interior, Fish and Wildlife Service, Washington, DC Resource Publication No. 153, PE17–215.

Hunter D (1953) Industrial toxicology. J Pharm Pharmacol, **5**:145–157.

HSDB (1994) Hazardous Substances Data Bank. National Library of Medicine, National Toxicology Information Program, Bethesda, MD, USA.

Ilivicky J, & Casida JE (1969) Uncoupling action of 2,4-dinitrophenols, 2-trifluoromethylbenzimidazols and certain other pesticide chemicals upon mitochondria from different sources and its relation to toxicity. Biochem Pharmacol, 18:1389–1401.

Ingebrigtsen K, & Froslie A (1980) Intestinal metabolism of DNOC and DNBP in the rat. Acta Pharmacol Toxicol, 46:326–328.

Izmerov NF, Sanotsky IV, & Sidorov KK (1982) Toxicometric parameters of industrial toxic chemicals under single exposure. Moscow, UNEP.

Jafvert CT (1990) Sorption of organic acid compounds to sediments: initial model development. Envir Toxicol Chem, 9:1259–1268.

Janda J (1970) On the toxicity of DNOC to pheasants, partridges and hares. Scien Agricult Bohem, 2(4):301–312.

Jastroch W, Knoll W, Lange B, Riemer F, & Thiele E (1978) [Results of a study on the exposure of agricultural workers to DNOC. Z Ges Hyg, 24:340–343 (in German).

Jegatheeswaran T, & Harvey DG (1970) The metabolism of DNOC in sheep. Vet Rec, 4 July, 19–20.

Jensen HL (1966) [Biological decomposition of herbicides in the soil. IV Dinitro-*ortho*-cresol.] Tidssk Planteal, 70:149–159 (in Danish).

Jensen HL, & Gundersen K (1955) Biological decomposition of aromatic nitro-compounds. Nature, 175:341.

Jonas W (1995) Adsorption/desorption of the test substance: [^{14}C]-DNOC Study no. NA 93 9714, NATEC Institut, Hamburg, Germany (unpublished report prepared for Elf Atochem Agri SA).

Jones GDG, & Edwards RA (1952) Studies of toxicity of 3,5-dinitro-*ortho*-cresol and its sodium salt to the honey bee. Bull Entomol Res, 43:67–78.

Jongerius O & Jongeneelen FJ (1991) Criteria document for an occupational exposure limit value of 4,6-dinitro-*o*-cresol (CAS 534–52–1). Commission of the European Communities, Directorate General Employment, Industrial Relations and Social Affairs, Health and Safety Directorate, Industrial Medicine and Hygiene Unit. SEG/CDO/29.1992, Luxembourg.

Judah JD (1952) Mode of action of the nitrophenols. Proc R Soc Med, 45:574.

Kadenczki L, Arpad Z, & Gardi I (1992) Column extraction of residues of several pesticides from fruits and vegetables: a simple multiresidue analysis methods. J Am Org Anal Chem Int, 75:53–61.

Kelly J (1995) DNOC: 13-week oral (dietary administration) range-finding study in the mouse. (Corning Hazelton project no. CHE 1151/8). Harrogate, UK (unpublished report prepared for Elf Atochem Agri SA).

Keplinger ML, Lanier GE, & Deichmann WB (1959) Effects of environmental temperature on the acute toxicity of a number of compounds in rats. Tox Appl Pharmacol, **1**:156–161.

King E, & Harvey DG (1953a) Some observations on the absorption and excretion of 4,6-dinitro-*o*-cresol (DNOC) I. Blood dinitro-*o*-cresol levels in the rat and the rabbit following different methods of absorption. Biochem J, **53**:185–195.

King E, & Harvey DG (1953b) Some observations on the absorption and excretion of 4,6-dinitro-*o*-cresol. Biochem J, **53**:196–200.

Kirkland DJ (1984) Study to evaluate the chromosome damaging potential of DNOC by its effects on the bone marrow cells of treated rats. (Pharmatox study no. PHM 6/RBM/AR/KF6) Hannover, Germany (unpublished report prepared for Pennwalt Holland bv).

Kirkland DJ (1986) Study to evaluate the chromosome damaging potential of 4,6-dinitro-*o*-cresol (DNOC) by its effects on the bone marrow cells of treated mice. (Microtest Research study no. PEN 1/MBM/KF27/MB1) York, UK (unpublished report prepared for Pennwalt Holland bv).

Kreczko S, Zwierz K, & Jaroszewicz K (1974) Glycoprotein biosynthesis by the guinea pig liver in chronic 4,6-dinitro-*o*-cresol poisoning. Acta Biol Acad Sci Hung, **25**(3):167–171.

Kühn R, Monika P, Pernak K-D, & Winter A (1989) Results of the harmful effects of water pollutants to *Daphnia magna* in the 21-day reproduction test. Water Res, **23**: 501–510.

Lawford DJ, King E, & Harvey DG (1954) On the metabolism of some aromatic nitro-compounds by different species of animals. Part II. The elimination of various nitro-compounds from the blood of different species of animals. J Pharm Pharmacol, **6**:619–624.

Leegwater DC, van der Greef J, & Bos KD (1982) Integrated studies on the metabolic fate of DNOC. II. Biotransformation in mammals. Med Fac Landbouww Rijksuniv Gent, **47**(1):401–408.

Leuenberger C, Czuczwa J, & Tremp J (1988) Nitrated phenols in rain: atmospheric occurrence of phototoxic pollutants. Chemosphere, **17**:511–515.

Lopez-Avila V, Bauer K, & Milanes J (1993) Evaluation of Soxhlet extraction procedure for extracting organic compounds from soils and sediments. J Am Org Anal Chem Int, **76**:864–880.

Malter A (1949) Intoxications survenues au cours de la fabrication d'un insecticide à base de dinitrocrésol. Arch Belg Med Soc Hyg, **7**:475–481.

Martin CN (1981) Study to determine the ability of DNOC to induce mutations to ouabain and 6-thioguanine resistance in mouse lymphoma L5178Y cells. (Microtest Research study no. PHARM 2/ML/JC/JGL) York, UK (unpublished report prepared for Pennwalt Holland bv).

Marzin D (1991a) Recherche de mutagénicité sur *Salmonella typhimurium his* − selon la technique de B.N. Ames sur le produit dinitro-*ortho*-cresol. (Institut Pasteur de Lille study no. IPL-R 910801). Lille, France (unpublished report prepared for Elf Atochem Agri SA).

Marzin D (1991b) Etude de mutagenèse au locus HPRT sur cellules V79 de hamster chinois (résistance à la 6-thioguanine) sur le produit dinitro-*ortho*-cresol (2-méthyl-4,6-dinitrophénol). (Institut Pasteur de Lille study no. IPL-R 910904). Lille, France (unpublished report for Elf Atochem Agri SA).

Marzin D (1991c) Etude de l'activité génotoxique par la technique du micronucléus chez la souris sur le produit dinitro-*ortho*-crésol. (Institut Pasteur de Lille study no. IPL-R 910804) Lille, France (unpublished report prepared for Elf Atochem Agri SA).

Marzin D (1991d) Etude de l'activité génotoxique du produit dinitro-*ortho*-crésol par la recherche d'aberrations chromosomiques par analyse de métaphases sur lymphocytes humains en culture. (Institut Pasteur de Lille study no. IPL-R 911010). Lille, France (unpublished report prepared for Elf Atochem Agri SA).

Mayer FL, & Ellersieck MR (1986) Manual of acute toxicity: Interpretation and database of 410 chemicals and 66 species of freshwater animals. US Department of the Interior, Fish and Wildlife Services, Publication No. 160, PE17–2LS, Washington, DC

McGirr JL, & Papworth DS (1953) Toxic hazards of the newer insecticides and herbicides. Vet Rec **65**(48):857–862.

Metcalf RL (1978) Insect control technology. In: Kirk-Othmer encyclopaedia of chemical technology. Vol. 13, 3rd ed. New York, John Wiley & Sons, p. 428.

Mogensen BB, & Spliid NH (1995) Pesticides in Danish watercourses: occurrence and effects. Chemosphere, **31**(8):3977–3990.

Moreland DE (1980) Effects of toxicants on oxidative phosphorylation. In: Hogdson E, & Guthrie FE eds. Introduction to biochemical toxicology. New York, Elsevier, pp. 245–260.

Morgan DP (1982) Recognition and management of pesticide poisonings. 3rd ed. Washington, DC, US Environmental Protection Agency (EPA-540/9–80–005).

Müller J, & Haberzetti R (1980). Mutagenicity of DNOC in *Drosophila melanogaster*. Arch Toxicol Suppl, **4**:59–61.

Nehéz M, Selypes A, & Paldy A (1977) [Examination of dinitro-*o*-cresol-containing fertilizer for mutagenic effects.] Egeszsegtudomany **21**:237–243 (in Hungarian).

Nehéz M, Selypes A, Paldy A, & Berencsi G (1978) The mutagenic effect of a dinitro-*o*-cresol containing pesticide on mice germ cells. Ecotox Environ Safety, 2:401–405.

Nehéz M, Paldy A, & Selypes A (1981) The teratogenicity and mutagenicity effects of dinitro-*o*-cresol-containing herbicide on laboratory mouse. Ecotox Environ Safety, **5**:38–44.

Nehéz M, Selypes A, Paldy A, Mezzag E, Berencsi G and Jarmay K (1982) The effects of five weeks treatment with dinitro-*o*-cresol or trifluralin containing pesticides on the germ cells of male mice. J Appl Toxicol, **2**(4):179–180.

Nehéz M, Selypes A, Mazzag E and Berencsi G (1984) Additional data on the mutagenic effect of dinitro-*o*-cresol containing herbicides. Ecotox Environ Safety, 8:75–79.

NIOSH (1978) Criteria for a recommended standard: Occupational exposure to dinitro-*ortho*-cresol. Report No.78–131, PB80–17870–159. Washington, DC, National Institute for Occupational Safety and Health.

NIOSH (1984) Manual of analytical methods. 2nd ed. Vol 5. Method no. S166 Washington, DC, National Institute for Occupational Safety and Health.

Nishimura N, Nishimura H, & Oshima H (1982) Survey on mutagenicity of pesticides by the *Salmonella* microsome test. J Aichi Med Univ Assoc, **10**(4):305–312.

Parker VH (1949) Method for routine estimation of 3:5-dinitro-*o*-cresol (DNOC) Analyst, **74**: 646–647.

Parker VH (1952) Enzymic reduction of 2,4-dinitrophenol by rat-tissue homogenates. Biochem J, **51**:363–370.

Parker VH, Barnes JM, & Denz FA (1951) Some observations on the toxic properties of 3,5-dinitro-*ortho*-cresol. Br J Industr Med, **8**:226–235.

Pollard AB, & Filbee JF (1951) Recovery after poisoning with di-nitro-*ortho*-cresol. Lancet, **II**:618–619.

Prost G, Vial R, & Tolot F (1973) Intoxication par le dinitro-*ortho*-crésol avec atteinte hépatique. Arch Mal Prof, **34**(9):556–557.

Quidet P (1975) Résultats des enquêtes en France en 1972, 1973, 1974 sur les causes de mortalités du gibier. Influence des pesticides et évaluation des risques selon la nature des produits. Phytoma, July:26–32.

Quinto I, DeMarinis E, Mallardo M, Arcucci A, Della Morte R, & Staiano N (1989) Effect of DNOC, ferbam and imidan exposure on mouse sperm morphology. Mutat Res, **224**:405–408.

Roseboom H, Wammes JIJ, & Wegman RCC (1981) Determination of nitrophenol derivatives in various crops by reversed-phase ion-pair high-performance liquid chromatography. Anal Chim Acta, **132**:195–199.

Sainsbury CR, Mullee DM, & Bartlett AJ (1995) Analytical DNOC and technical DNOC. Characterisation (Safepharm Laboratories study no. 764/029) Derby,UK Safepharm Analytical Chemistry Laboratory (unpublished report prepared for Elf Atochem Agri SA).

Sewell IG, Mead C, & Bartlett AJ (1995a) Technical DNOC. Algal inhibition test. Safepharm Laboratories study no. 764/15. Derby, UK (unpublished report prepared for Elf Atochem Agri SA).

Sewell IG, Foulger J, & Bartlett AJ (1995b) Technical DNOC. Acute toxicity to bluegill sunfish (*Lepomis macrochirus*). Safepharm Laboratories study no. 764/12. Derby, UK (unpublished report prepared for Elf Atochem Agri SA).

Sewell IG, Foulger J, & Bartlett AJ (1995c) Technical DNOC: Acute toxicity to Rainbow trout (*Onchorhynchus mykiss*). Safepharm Laboratories study no. 764/13. Derby, UK (unpublished report prepared for Elf Atochem Agri SA).

Smith JN, Smithies RH, & Williams RT (1953) Studies in detoxication. 48. Urinary metabolites of 4,6-dinitro-o-cresol in the rabbit. Biochemistry, 54:224–230.

Somani SM, Schaeffer DJ, & Mack JO (1981) Quantifying the toxic and mutagenic activity of complex mixtures with *Salmonella typhimurium*. J Tox Environ Health, 7:643–653.

Spencer HC, Rowe VK, Adams EM, & Irish DD (1948) Toxicological studies on laboratory animals of certain alkyldinitrophenols used in agriculture. J Industr Hyg Tox, 30(1):10–25.

Starek A, & Lepiarz W (1974) Influence of some fats on 4,6-dinitro-o-cresol (DNOC) resorption from alimentation tract of rats. Pol J Pharmacol Pharm, 26:485–491.

Steer C (1951) Death from di-nitro-*ortho*-cresol. Lancet, I:1419.

Stott H (1953) The use of certain toxic chemicals in agricultural practice. East African Med J, 30:59–67.

Stott H (1956) Polyneuritis after exposure to dinitro-*ortho*-cresol. BMJ, 1:900–901.

Sundvall A, Marklund H, & Rannung U (1984) The mutagenicity of *Salmonella typhimium* of nitrobenzoic acids and other wastewater components generated in the production of nitrobenzoic acids and nitrotoluenes. Mutat Res 137:71–78.

Tan GH, & Chong CL (1993) Trace monitoring of water-borne phenolics in the Klang river basin. Environ Monit Assess, 24:267–277.

Tesic D, Terzic LJ, Dimitrijevic B, Zivanov D, & Slavic M (1972) Experimental investigation on the effect of environmental temperature and some medicaments on the toxic effect of dinitroorthocresol. Acta Vet Beograd, 22(2):45–52.

Tewfik MS, & Evans WC (1966) The metabolism of 3,5-dinitro-o-cresol (DNOC) by soil micro-organisms. Biochem J, 99:31–32.

Til HP (1980) Sub-chronic (90-day) oral toxicity study with DNOC in dogs. CIVO TNO Institute study no. B 80/0359, Zeist, The Netherlands (unpublished report prepared for Pennwalt Holland bv).

Til HP, & Kengen MTF (1980) Subacute (8-day) dietary LC50 study with DNOC in Japanese quail. CIVO-TNO study no. R 6596. CIVO-TNO, Zeist, The Netherlands. (unpublished report prepared for Pennwalt Holland bv).

Tomlin C ed. (1997) The pesticide manual, a word compendium. 11th ed. Farnham, UK, British Crop Protection Council.

Tremain SP, & Bartlett AJ (1995) Technical DNOC (batch number 547–108). Determination of hazardous physico-chemical properties. (Safepharm Laboratories study no. 764/003) Derby, UK Safepharm Analytical Chemistry Laboratory (unpublished report prepared for Elf Atochem Agri SA).

Tremp J, Mattrei P, & Giger W (1993) Phenols and nitrophenols as tropospheric pollutants: Emissions from automobile exhausts and phase transfer in the atmosphere. Water, Air Soil Pollution, **68**:113-123.

Tripathi AM, Mhalas JG, & Rama-Rao NV (1989) Determination of 2,6 and 4,6-dinitrocresols by high performance liquid chromatography on a betacyclodextrin bonded column. J Chromatogr, **466**:442-445.

Truhaut R & de Lavaur E (1967) Etude du métabolisme du dinitro-4,6-orthocrésol chez le lapin. C R Acad Sci Paris Ser D, **264**:1938-1940.

UNEP Chemicals (IRPTC) (1999) UNEP Chemicals data profile (legal) on 4,6 DNOC. Geneva, UNEP Chemicals.

US EPA (1984a) Method 604-phenols. Environmental Protection Agency. Code of Federal Regulations. 40 CFR 136.

US EPA (1984b) Method 625-Base/Neutrals and Acids. Environmental Protection Agency. Code of Federal Regulations. 40 CFR 136.

US EPA (1984c) Method 1625 Revision B-Semivolatile Organic Compounds by Isotope Dilution GC/MS. Environmental Protection Agency. Code of Federal Regulations. 40 CFR 136

US EPA (1986a) Phenols-method 8040. In: Test methods for evaluating solid wastes. SW-846. 3rd ed. Volume 1B: Laboratory Manual. Washington, DC: US Environmental Protection Agency, Office of Solid Waste and Emergency Response.

US EPA (1986b) Gas chromatography/mass spectrometry for semivolatile organics: capillary column technique – method 8270. In: Test methods for evaluating solid wastes. SW-846. 3rd ed. Volume 1B: Laboratory Manual. Washington, DC: US Environmental Protection Agency, Office of Solid Waste and Emergency Response.

US EPA (1988) Health and environmental effects profile for dinitrocresols. Cincinnati, OH: US Environmental Protection Agency. NTIS PB88-220769.

US EPA (1993) Status of pesticides in reregistration and special review. EPA 738/R/93/009. Washington, DC: Office of Prevention, Pesticides and Toxic Substances.

van de Berg KJ, van Raaij JAGM, Bragt PC, & Notten WRF (1991) Interactions of halogenated industrial chemicals with transthyretin and effects on thyroid hormone levels in vivo. Arch Toxicol, **65**:15-19.

van der Greef J, & Leegwater DC (1983) Urine profile analysis by field desorption mass spectrophotometry, a technique for detecting metabolites of xenobiotics. Application to 3,5-dinitro-2-hydroxytoluene (DNOC). Biomed Mass Spectrom, **10**:1-4.

van der Hoeven JCM (1984) Assessment of the effects of 4,6-dinitro-*o*-cresol (DNOC) on the reproduction of *Daphnia magna* (Notox Toxicological Research and Consultancy, s'Hertogenbosch, The Netherlands) (unpublished report prepared for Pennwalt Holland bv).

References

van der Hoeven JCM (1992) Acute toxicity of DNOC technical to the worm species *Eisenia fetida*. TNO study no. R91/324. TNO Institute of Delft, The Netherlands (unpublished report prepared for Elf Atochem Agri SA).

van der Laar RTH, de Vries I, & Meulenbelt J (1993) [Acute occupational intoxications through the use of pesticides in forestry, agriculture and horticulture.]. National Poison Control Center of the National Institute of Public Health and Environmental Protection, The Hague (in Dutch).

van Noort HR (1960) [Dinitro-*ortho*-cresol intoxication in sprayers.] Nederlands tijdschrift voor geneeskunde 104:676–684 (in Dutch).

Verheij ER, & van der Graaf J (1995) The identification of a DNOC metabolite in soil. Study no. 274514, TNO Nutrition and Food Research, Zeist, The Netherlands. (unpublished report prepared for Elf Atochem Agri bv).

Vonk JW, & van der Hoven A (1981) Degradation and conversion of DNOC in water. Adsorption of DNOC to bottom sludge. TNO Organisch Chemisch Intituut, The Netherlands (unpublished report prepared for Elf Atochem Agri bv).

Vos JG, Krajnc EI, Beekhof PK, & van Logten MJ (1983) Methods for testing immune effects of toxic chemicals: evaluation of the immunotoxicity of various pesticides in the rat. In: Miyamoto J ed. IUPAC Pesticide chemistry: human welfare and the environment. Kyoto, Japan, Pergamon Press, pp. 497–504.

Wodageneh A (1997) Obsolete and unwanted pesticide stocks. In: Prevention and disposal of obsolete and unwanted pesticide stocks in Africa and the Near East. Second Consultation Meeting. FAO Pesticide Series, 5:5–12.

WHO (1982) Recommended health limits in occupational exposure to pesticides. Technical Report Series 677. Geneva, World Health Organization.

WHO (1999) The WHO recommended classification of pesticide by hazards and guidelines to classification 1998–1999. WHO/PCS/98.21. Geneva, World Health Organization.

Yao S, Meyer A, & Henze G (1991) Comparison of amperometric and UV-spectrophotometric monitoring in HPLC analysis of pesticides. Fresenius J Anal Chem, **339**:207–211.

RESUME ET CONCLUSIONS

Résumé

.1 Identité, propriétés physiques et chimiques et méthodes d'analyse

Le DNCO (4,6-dinitro-orthocrésol) se présente sous la forme d'un solide cristallin jaunâtre. Son point de fusion est de 88,2-88,9 °C et sa tension de vapeur est égale à $1,6 \times 10^{-2}$ Pa à 25°C. Sa solubilité dans l'eau est de 6,94 g/litre à 20°C et à pH 7 et elle dépend largement du pH. Dans l'eau stérile, le DNOC est relativement stable. Les méthodes d'analyse utilisées pour doser le DNOC dans des échantillons provenant dans l'environnement sont les suivantes : chromatographie en phase liquide à haute performance avec détection par absorption UV (CLHP/UV)ou chromatographie en phase gazeuse avec détecteur azote - phosphore (CPG/NP), par ionisation de flamme ou par couplage avec un spectromètre de masse. Dans les liquides biologiques, le dosage s'effectue habituellement par spectrophotométrie ou en ayant recours à des techniques plus récentes (CPG/NP ou CLHP/UV).

.2 Sources d'exposition humaine et environnementale

Le DNOC s'emploie en agriculture comme larvicide, ovicide ou insecticide (contre les sauterelles et autres insectes) ainsi que pour provoquer la dessiccation des fanes de pomme de terre. On l'utilise également comme inhibiteur de polymérisation et comme intermédiaire dans l'industrie chimique. Les formulations de DNOC à usage agricole sont des concentrés émulsionnables aqueux ou huileux.

.3 Transport, distribution et transformation dans l'environnement

Dans les eaux superficielles, le DNOC a une demi-vie de 3 à 5 semaines. Etant donné sa faible tension de vapeur et sa solubilité modérée dans l'eau, il n'a aucune tendance à s'évaporer. Dans le sol, le DNOC est rapidement décomposé par les microorganismes avec une durée médiane de décomposition (TD_{50}) qui se situe entre 1,7 et 15 jours. On en a identifié quelques métabolites environnementaux, qui résultent d'une biotransformation par réduction, peut-être suivie d'une dégradation plus poussée par oxydation. Le DNOC non dissocié est fortement adsorbé aux particules du sol lorsque le pH est faible, mais la sorption est limitée aux valeurs du pH rencontrées dans l'environnement. En pratique, on ne constate guère de lessivage vers les eaux souterraines, probablement par suite de la biodégradation du composé.

1.4 Niveaux dans l'environnement et exposition humaine

Les principales sources d'exposition humaine sont les contacts qui peuvent se produire pendant la fabrication, l'utilisation en agriculture ou dans l'industrie des matières plastiques. Comme la forte toxicité aiguë du produit est bien connue et qu'il a tendance à fortement colorer la peau en jaune, les ouvriers agricoles n'hésitent pas à revêtir une tenue protectrice pour exposer leur épiderme le moins possible. Le DNOC est souvent présenté et transporté sous la forme d'une poudre humidifiée (12 % d'eau en poids) afin de réduire le risque d'exposition aux poussières. On peut s'attendre à des cas d'exposition professionnelle en agriculture et dans l'industrie chimique.

1.5 Cinétique et métabolisme

Le métabolisme du DNOC est de même nature chez un certain nombre d'espèces. En revanche, sa vitesse d'élimination varie notablement d'une espèce à l'autre. Ainsi, il subsiste plus longtemps chez l'Homme que chez les animaux de laboratoire.

1.6 Effets sur les mammifères de laboratoire et les systèmes d'épreuve in vitro

1.6.1 Exposition unique

La dose létale médiane par voie orale (DL_{50}) varie de 20 à 85 mg/kg de poids corporel chez le rat et de 50 à 100 mg/kg p.c. chez le porc. Par voie percutanée, la DL_{50} se situe entre 600 mg/kg et plus de 2000 mg/kg p.c. chez le rat ; elle est de 1000 mg/kg p.c. chez le lapin. Ces valeurs montrent que la résorption cutanée est limitée. La concentration létale médiane par inhalation (CL_{50}) a été trouvée égale à 230 mg/m^3 pour une exposition de 4h chez le rat et à 40 mg/m^3 pour une exposition de même durée chez le chat.

1.6.2 Exposition de brève durée

En administrant pendant une brève durée (90 jours au maximum) du DNOC à des rats, des souris et des chiens en mélange avec leur nourriture, on a constaté une réduction du poids corporel sans modification notable de la consommation de nourriture. A forte dose, des effets ont été constatés au niveau du foie, comme en témoignait l'augmentation de l'activité des enzymes hépatiques. On constatait également une augmentation de l'urée sanguine.

1.6.3 Irritation cutanée et oculaire et sensibilisation cutanée

L'application de DNOC sur la peau de lapins a produit un érythème et un oedème, ce qui traduit un effet irritant. Chez le cobaye, le DNOC produit une sensibilisation cutanée et il se montre agressif pour la muqueuse oculaire chez le lapin.

1.6.4 Exposition de longue durée

Lors d'une étude d'alimentation de longue durée sur des rats, on a constaté que le DNOC ne produisait pas d'effets indésirables attribuables à sa présence à des doses quotidiennes allant jusqu'à 5 mg/kg p.c. On a observé que dans le groupe soumis à la dose la plus élevée, la consommation de nourriture était légèrement plus forte (+ 6%) que dans le groupe témoin. Cet effet (consommation de nourriture augmentée sans conséquence pour le gain de poids) résulte du mode d'action particulier du DNOC.

1.6.5 Reproduction, embryotoxicité et tératogénicité

A forte dose, le DNOC a un léger effet sur la reproduction qui se manifeste par une réduction du poids corporel et de la taille des portée. Les autres paramètres génésiques n'accusent aucune modification. Le DNOC n'a eu aucun effet tératogène chez des rattes gravides qui en avaient reçu par voie orale des doses quotidiennes allant jusqu'à 25 mg/kg p.c. du jour 6 au jour 15 de la gestation inclusivement. Chez des lapins, également traités par voie orale, une dose élevée (25 mg/ kg p.c. par jour) a provoqué des intoxications mortelles chez les mères. A cette dose, on a observé des effets tératogènes, notamment des micro-ophtalmies ou des anophtalmies, des hydrocéphalies ou des microcéphalies. Administré par application cutanée à des lapines gravides, le DNOC s'est révélé toxique pour elles à dose élevée (90 mg/kg p.c. par jour) et a eu quelques effets embryotoxiques, à l'exclusion de toute tératogénicité. On n'a relevé aucun effet tératogène ou embryotoxique chez des souris gravides traitées par voie orale ou intrapéritonéale.

1.6.6 Mutagénicité

Les données disponibles restent ambiguës quant au pouvoir mutagène du DNOC.

1.6.7 Cancérogénicité

Lors d'une étude d'alimentation de longue durée sur des rats, on n'a pas observé d'augmentation de l'incidence des tumeurs – quel qu'en soit le type- qui soit imputable à la présence de DNOC.

1.7 Effets sur l'Homme

Des cas d'intoxication aiguë par le DNOC ont été observés chez l'Homme. Les symptômes de cette intoxication sont les suivants : agitation, sensation de chaleur, rougeurs, sudation, soif, respiration profonde et rapide, tachycardie, forte augmentation de la température centrale et cyanose aboutissant à un collapsus, au coma et à la mort. Une température ambiante élevée accroît encore ces effets, qui correspondent bien au mécanisme proposé pour expliquer l'action du DNOC.

1.8 Effets sur les êtres vivants dans leur milieu naturel

Au doses qui sont recommandées pour son épandage, le DNOC n'a guère d'effets sur les microorganismes terricoles. Sa toxicité aiguë pour les organismes aquatiques est très variable, même parmi les animaux dont la valeur de la CL_{50} est comprise entre 0,07 et 5,7 mg/litre. Lors des tests pratiqués en laboratoire, ce sont les poissons qui se sont révélés les plus sensibles. Le calcul du rapport toxicité/exposition (TER) montre que pour les organismes aquatiques, il existe un risque dû aux embruns. Le respect d'une zone de sécurité de 5 m ramène le risque à un niveau acceptable. Le DNOC est fortement toxique pour les abeilles mais il est probable que celles-ci sont peu exposées ; le quotient de dangerosité pour les abeilles indique que le risque est faible. Dans le cas des lombrics, le TER (CL_{50} de 17 mg/kg de terre) montre que l'épandage de DNOC comme desséchant ne comporte qu'un risque modéré. La forte toxicité aiguë du DNOC pour les oiseaux et les mammifères ne devrait pas se manifester dans l'environnement car l'exposition est vraisemblablement faible. Des observations limitées effectuées sur le terrain corroborent cette assertion. Il n'est guère possible de caractériser le risque de manière plus précise car les données sur les résidus et les effets obtenues en situation réelle font défaut.

Conclusions

Si l'on se conforme aux recommandations officielles et que l'on prenne les mesures de protection individuelle habituelles, l'exposition au DNOC est réduite dans une forte proportion et ramenée à des niveaux où il n'y a plus de risque d'intoxication générale. Les modalités actuelles d'utilisation des produits phytosanitaires à base de DNOC ne donnent lieu à aucun résidu décelable dans les cultures traitées ; il n'y a donc pas d'exposition de la population dans son ensemble. Le DNOC provoque une sensibilisation cutanée chez le cobaye. Le calcul du facteur de risque résultant de son utilisation comme desséchant ou en épandage sur les arbres fruitiers en période de dormance montre qu'il y a possibilité d'effets nocifs pour les organismes aquatiques (par suite de la dispersion des embruns) et les lombrics. Dans leur milieu naturel, les autres êtres vivants ne courent vraisemblablement guère de risque , du fait que l'exposition est faible. On n'a pas cherché à évaluer les risques découlant des autres usages du DNOC (par exemple pour combattre les sauterelles) car on manque de données sur les doses d'emploi et les méthodes d'épandage.

RESUMEN Y CONCLUSIONES

I Resumen

1.1 Identidad, propiedades físicas y químicas y métodos analíticos

El DNOC (4-6 dinitro-orto-cresol) es un sólido cristalino de color amarillento, cuyo punto de fusión es 88,2°C-88,9 °C y su presión de vapor $1,6 \times 10^{-2}$ a 25 °C. La solubilidad del DNOC en agua es 6,94 g/litro a 20 °C y pH 7, y depende en gran medida del pH. El DNOC es un producto relativamente estable en agua estéril. Su análisis en el medio ambiente se realiza mediante cromatografía líquida de alto rendimiento (CLAR) con detección ultravioleta (UV) o por cromatografía de gases (CG) con detección de nitrógeno-fósforo (DNF), detección por ionización de llama o espectrometría de masas. En los líquidos biológicos, la determinación del DNOC se suele realizar mediante espectrofotometría y más recientemente mediante CG/DNF o CLAR/UV.

1.2 Fuentes de exposición humana y ambiental

El DNOC se utiliza en la agricultura como larvicida, ovicida e insecticida (contra la langosta y otros insectos) y como desecante de las matas de papa. Se emplea también como inhibidor de la polimerización y como intermediario en la industria química. Para usos agrícolas, el DNOC se formula principalmente como concentrado emulsionable, acuoso o bien oleoso.

1.3 Transporte, distribución y transformación en el medio ambiente

La semivida del DNOC en el agua superficial es de 3-5 semanas. Su baja presión de vapor y su moderada solubilidad en agua indican que no tiene capacidad para volatilizarse. En el suelo, los microorganismos lo degradan con rapidez, con valores del tiempo de degradación mediano (TD_{50}) del orden de 1,7-15 días. Se han identificado varios metabolitos en el medio ambiente, procedentes de una transformación reductora posiblemente seguida de una ulterior degradación oxidativa. Con un pH bajo, el DNOC no disociado se adsorbe sobre las partículas mediante una unión fuerte, pero la sorción está limitada al pH apropiado en el medio ambiente. En la práctica, apenas se ha detectado lixiviación al agua freática, probablemente debido a la biodegradación.

1.4 Niveles en el medio ambiente y exposición humana

La principales fuentes de exposición humana son el contacto durante la fabricación y el uso en la agricultura y en la industria de los plásticos. Debido a su conocida toxicidad aguda y a que tiñe de un color amarillento intenso la piel, los agricultores tienen el cuidado de utilizar ropa de protección adecuada a fin de reducir la exposición cutánea. En la industria de los plásticos, el DNOC se fabrica y transporta como polvo, con frecuencia amortiguado con agua (12% en peso) para reducir el riesgo de exposición de los trabajadores al polvo. Cabe prever exposición ocupacional en la agricultura y la industria química.

1.5 Cinética y metabolismo

La ruta metabólica del DNOC es cualitativamente semejante en varias especies. Sin embargo, su tasa de eliminación varía de manera considerable de unas a otras. Las personas retienen el DNOC más tiempo que otras especies sometidas a prueba.

1.6 Efectos en mamíferos de laboratorio; sistemas de prueba in vitro

1.6.1 Exposición única

Los valores de la dosis letal mediana (DL_{50}) del DNOC por vía oral oscilan entre 20 y 85 mg/kg de peso corporal en la rata y entre 50 y 100 mg/kg de peso corporal en el cerdo. Su DL_{50} por vía percutánea varía entre 600 y más de 2000 mg/kg de peso corporal en el conejo, lo que indica una absorción cutánea limitada. Se han determinado unos valores de la concentración letal mediana (CL_{50}) por inhalación de 230 mg/m^3 de peso corporal para una exposición de cuatro horas en la rata y de 40 mg/m^3 (4 horas) en el gato.

.6.2   Exposición breve

La administración de DNOC con los alimentos durante un período breve de hasta 90 días produjo una reducción del aumento del peso corporal en ratas, ratones y perros, normalmente sin alteración significativa del consumo de alimentos. Con dosis más elevadas se vio afectado el hígado, como puso de manifiesto una mayor actividad de las enzimas hepáticas. Con dosis altas también aumentó la concentración de urea en sangre.

.6.3   Irritación cutánea y ocular y sensibilización cutánea

La aplicación cutánea de DNOC produjo en los conejos eritema y edema, lo cual indica un efecto irritante. El DNOC es sensibilizador cutáneo en el cobaya y tiene un efecto corrosivo en los ojos del conejo.

.6.4   Exposición prolongada

En un estudio con ratas en el que se les suministró DNOC con los alimentos durante un período prolongado, no se detectaron efectos adversos relacionados con el tratamiento utilizando dosis de hasta 5 mg/kg de peso corporal al día. Se observó que el consumo de alimentos en el grupo que recibía la dosis más elevada era ligeramente superior (+6%) al testigo no tratado. Este efecto (es decir, mayor consumo de alimentos sin efecto en el aumento del peso corporal) es una consecuencia del mecanismo de acción específico del producto.

.6.5   Reproducción, embriotoxicidad y teratogenicidad

Con dosis elevadas, el DNOC tiene un efecto ligero en la reproducción en forma de reducción del peso corporal y del tamaño de la camada. No afecta a otros parámetros de la reproducción. No indujo efectos teratogénicos en ratas preñadas que recibieron por vía oral dosis de hasta 25 mg/kg de peso corporal al día desde el sexto día de gestación hasta el 15 inclusive. El suministro a conejos de un tratamiento por vía oral con dosis altas de 25 mg/kg de peso corporal al día produjo toxicidad materna y fue causa de mortalidad. Con estas dosis se detectaron efectos teratogénicos, en particular microftalmia o anoftalmia e hidrocefalia o microcefalia. Cuando se administró a conejas por vía cutánea durante la gestación, indujo toxicidad materna con una dosis elevada de 90 mg/kg de peso corporal al día, que dio lugar a una cierta embriotoxicidad, pero no se detectó teratogenicidad. No se registraron pruebas de teratogenicidad o embriotoxicidad en ratones tratados por vía oral o intraperitoneal durante la gestación.

1.6.6   *Mutagenicidad*

Sobre la base de todos los datos disponibles, la mutagenicidad del DNOC sigue siendo equívoca.

1.6.7   *Carcinogenicidad*

En un estudio de administración prolongada con los alimentos a ratas, el DNOC no provocó un aumento de la incidencia de ningún tipo de tumor.

1.7   *Efectos en el ser humano*

El DNOC ha producido intoxicaciones agudas en el ser humano. Los síntomas asociados con su toxicidad son agitación, sensación de calor, enrojecimiento de la piel, sudoración, sed, respiración profunda y rápida, taquicardia, fuerte aumento de la temperatura corporal y cianosis que lleva al colapso, el coma y la muerte. Los efectos se ven potenciados por una temperatura ambiental elevada. Estos efectos están en consonancia con el mecanismo de acción propuesto del DNOC

1.8   *Efectos en los seres vivos del medio ambiente*

El DNOC tiene pocos efectos en los microorganismos del suelo con las dosis de aplicación recomendadas. La toxicidad aguda para los organismos acuáticos es muy variable, incluso dentro de los distintos grupos de animales, con valores de la CL_{50} que oscilan entre 0,07 y 5,7 mg/litro; los peces fueron las especies más sensibles en las pruebas de laboratorio. Las razones toxicidad:exposición calculadas para los organismos acuáticos indican cierto riesgo derivado de la tendencia a la pulverización. La utilización de una zona de protección de cinco metros reduce los factores de riesgo a niveles aceptables. El DNOC tiene una toxicidad aguda para la abeja melífera, pero probablemente la exposición es baja; los cocientes de peligro para estas abejas indican un riesgo bajo. La razón toxicidad:exposición para la lombriz de tierra (CL_{50} con 17 mg/kg de suelo) pone de manifiesto un riesgo moderado tras el uso de DNOC como desecante. No es probable que el DNOC produzca una toxicidad aguda alta en las aves y los mamíferos en el medio ambiente, porque la exposición probablemente es baja. Esta conclusión está respaldada por informes limitados de incidentes sobre el terreno. No es posible caracterizar el riesgo con más detalle, porque no se dispone de información relativa a los residuos y sus efectos en el medio ambiente.

Conclusiones

Cuando se utiliza de acuerdo con las recomendaciones registradas, junto con las medidas habituales de protección individual, la exposición al DNOC se reduce enormemente, quedando en niveles que no provocan toxicidad sistémica. Habida cuenta de las pautas actuales de uso de productos de protección fitosanitaria que contienen DNOC como principio activo, no hay residuos detectables en los cultivos tratados, de manera que no hay exposición de la población general. El DNOC es sensibilizador cutáneo en los cobayas. El uso agrícola como desecante y en los cultivos de frutas en fase de latencia da lugar a factores de riesgo calculados que indican posibles efectos adversos en los organismos acuáticos (debido a la tendencia a la pulverización) y las lombrices de tierra. No es probable que produzca efectos adversos en otros organismos de la naturaleza, porque la exposición será baja. No se ha intentado una evaluación del riesgo para otros posibles usos del DNOC (por ejemplo, en la lucha contra la langosta), debido a la falta de información sobre las tasas y los métodos de aplicación.

THE ENVIRONMENTAL HEALTH CRITERIA SERIES (continued)

Flame retardants: tris(2-butoxyethyl) phosphate, tris(2-ethylhexyl) phosphate and tetrakis(hydroxymethyl) phosphonium salts (No. 218, 2000)
Fluorine and fluorides (No. 36, 1984)
Food additives and contaminants in food, principles for the safety assessment of (No. 70, 1987)
Formaldehyde (No. 89, 1989)
Fumonisin B_1 (No. 219, 2000)
Genetic effects in human populations, guidelines for the study of (No. 46, 1985)
Glyphosate (No. 159, 1994)
Guidance values for human exposure limits (No. 170, 1994)
Heptachlor (No. 38, 1984)
Hexachlorobenzene (No. 195, 1997)
Hexachlorobutadiene (No. 156, 1994)
Alpha- and beta-hexachlorocyclohexanes (No. 123, 1992)
Hexachlorocyclopentadiene (No. 120, 1991)
n-Hexane (No. 122, 1991)
Human exposure assessment (No. 214, 2000)
Hydrazine (No. 68, 1987)
Hydrogen sulfide (No. 19, 1981)
Hydroquinone (No. 157, 1994)
Immunotoxicity associated with exposure to chemicals, principles and methods for assessment (No. 180, 1996)
Infancy and early childhood, principles for evaluating health risks from chemicals during (No. 59, 1986)
Isobenzan (No. 129, 1991)
Isophorone (No. 174, 1995)
Kelevan (No. 66, 1986)
Lasers and optical radiation (No. 23, 1982)
Lead (No. 3, 1977)[a]
Lead, inorganic (No. 165, 1995)
Lead – environmental aspects (No. 85, 1989)
Lindane (No. 124, 1991)
Linear alkylbenzene sulfonates and related compounds (No. 169, 1996)
Magnetic fields (No. 69, 1987)
Man-made mineral fibres (No. 77, 1988)
Manganese (No. 17, 1981)
Mercury (No. 1, 1976)[a]
Mercury – environmental aspects (No. 86, 1989)
Mercury, inorganic (No. 118, 1991)
Methanol (No. 196, 1997)
Methomyl (No. 178, 1996)
2-Methoxyethanol, 2-ethoxyethanol, and their acetates (No. 115, 1993)
Methyl bromide (No. 166, 1995)
Methylene chloride
 (No. 32, 1984, 1st edition)
 (No. 164, 1996, 2nd edition)
Methyl ethyl ketone (No. 143, 1992)
Methyl isobutyl ketone (No. 117, 1990)
Methylmercury (No. 101, 1990)

Methyl parathion (No. 145, 1992)
Methyl *tertiary*-butyl ether (No. 206, 1998)
Mirex (No. 44, 1984)
Morpholine (No. 179, 1996)
Mutagenic and carcinogenic chemicals, guide to short-term tests for detecting (No. 51, 1985)
Mycotoxins (No. 11, 1979)
Mycotoxins, selected: ochratoxins, trichothecenes, ergot (No. 105, 1990)
Nephrotoxicity associated with exposure to chemicals, principles and methods for the assessment of (No. 119, 1991)
Neurotoxicity associated with exposure to chemicals, principles and methods for the assessment of (No. 60, 1986)
Nickel (No. 108, 1991)
Nitrates, nitrites, and N-nitroso compounds (No. 5, 1978)[a]
Nitrogen oxides
 (No. 4, 1977, 1st edition)[a]
 (No. 188, 1997, 2nd edition)
2-Nitropropane (No. 138, 1992)
Noise (No. 12, 1980)[a]
Organophosphorus insecticides: a general introduction (No. 63, 1986)
Paraquat and diquat (No. 39, 1984)
Pentachlorophenol (No. 71, 1987)
Permethrin (No. 94, 1990)
Pesticide residues in food, principles for the toxicological assessment of (No. 104, 1990)
Petroleum products, selected (No. 20, 1982)
Phenol (No. 161, 1994)
d-Phenothrin (No. 96, 1990)
Phosgene (No. 193, 1997)
Phosphine and selected metal phosphides (No. 73, 1988)
Photochemical oxidants (No. 7, 1978)
Platinum (No. 125, 1991)
Polybrominated biphenyls (No. 152, 1994)
Polybrominated dibenzo-*p*-dioxins and dibenzofurans (No. 205, 1998)
Polychlorinated biphenyls and terphenyls
 (No. 2, 1976, 1st edition)[a]
 (No. 140, 1992, 2nd edition)
Polychlorinated dibenzo-*p*-dioxins and dibenzofurans (No. 88, 1989)
Polycyclic aromatic hydrocarbons, selected non-heterocyclic (No. 202, 1998)
Progeny, principles for evaluating health risks associated with exposure to chemicals during pregnancy (No. 30, 1984)
1-Propanol (No. 102, 1990)
2-Propanol (No. 103, 1990)
Propachlor (No. 147, 1993)
Propylene oxide (No. 56, 1985)
Pyrrolizidine alkaloids (No. 80, 1988)
Quintozene (No. 41, 1984)
Quality management for chemical safety testing (No. 141, 1992)
Radiofrequency and microwaves (No. 16, 1981)

---

[a] Out of print

www.ingramcontent.com/pod-product-compliance
Ingram Content Group UK Ltd.
Pitfield, Milton Keynes, MK11 3LW, UK
UKHW021309180426
11947UKWH00015B/1111